HUMBOLDT JOURNAL OF MICROBIOLOGY

Volume 24, 2025

Microbiology Laboratory
Cal Poly Humboldt
Arcata, CA

THE PRESS AT CAL POLY HUMBOLDT

The Press at Cal Poly Humboldt

Cal Poly Humboldt Library
1 Harpst St.
Arcata, CA 95524-8299

press.humboldt.edu

ISBN: 978-1-962081-41-2

Layout & Design by Mia Costales

Cover photo by student authors
Background cover image fom Adobe Stock

The Press at Cal Poly Humboldt publishes high-quality scholarly, intellectual, and creative works by or in support of our campus community. The press supports the Cal Poly Humboldt purpose to improve the human condition and our environment by promoting understanding of social, economic, and environmentcl issues.

TABLE OF CONTENTS

5 Comparative Analysis of Taproot and Fibrous Root System Microbiomes Under Flooding Stress
Natalie Kozlowski, Leighanna Jake

11 Elucidating the Relationship Between Water Quality and Antibiotic Resistance of Rainwater Microbes Across Western Humboldt County
Theo Murphy, Eve Wendley, Justice Laskowski, Tyler Paredes

20 Use of ***Enterococcus faecalis*** and Cetylpyridinium Chloride in Altering Canine Mouth Microbiome
Curtis Cline, Quigley Espinola, Sofia Lopez-Orenday, Scarlet Renner, Alex Yasumatsu

26 Growth Trends in Abundance and Diversity of Synthetic Microbial Communities over Varying Thermal Conditions
Mariska Kessler, Jocelyn Wolfinger, Destiny Alcaraz

36 Comparative Analysis of Probiotic Viability in Alcoholic and Non-Alcoholic Kombucha
Isabella Cerrone, Chloe Kraft, Brooke Pirkle, Gianna Vendrell, Danielle Williamson

42 Solvent extracts of ***Pulvigera papillosa*** (Orthotrichaceae) show no antimicrobial properties in disk diffusion tests
Justin Paulin, Michael Lanier, Ryan Solorzano, Grayson Prater

47 Assessing Antimicrobial Properties of Plant-derived Ellagic Acid Supplements
Noah Schuhmann, Andrew McLaughlin, Julian Barrera

Comparative Analysis of Taproot and Fibrous Root System Microbiomes Under Flooding Stress

Natalie Kozlowski, Leighanna Jake

ABSTRACT

There has been increased interest in understanding how the microbiomes of vascular plants respond to flooding stress and many studies have focused on monocotyledonous plants with fibrous root systems. This leaves a gap in knowledge of dicotyledonous plants with taproot systems, as previous studies have been in the context of flooding to cropland affecting major food crops which are often monocotyledonous plants with fibrous root systems. Since the structural and functional differences between these root systems may influence microbial community stability and response to flooding conditions, a study looking at both is long overdue. This study aims to investigate whether microbial communities associated with taproot systems found in dicotyledonous plants have different responses to flooding stress compared to those of previously documented fibrous root systems. To investigate this, we cultivated *Lolium multiflorum* and *Brassica rapa* and submitted them to flooding stress taking samples from the roots and rhizosphere in three instances: before flooding, after one day of flooding, and after 3 days of flooding. The results show a statistically significant difference in the richness and species composition of microorganisms between root types. A greater understanding of how the microbial communities of different root structures respond to flooding is key to understanding the agricultural and ecological effects of flooding.

INTRODUCTION

The relationship between humans and plants is extensive, with more than half of the world's population relying on agricultural grains for the majority of their caloric intake (1). Major cereal crops worldwide are almost all monocots and therefore have fibrous root systems and are consistently at risk from abiotic environmental factors such as flooding (2). Flooding reduces the fitness and production of plants through reduced access to oxygen for the plants and mutualists in the soil (3). As cropland exposure to flooding increases each year across the globe, so does the need to understand the effects of flooding and oxygen loss plants and their microbial communities in all plant types (4).

The relationship between plants and the microscopic world is most apparent when looking at the complex communities supported by plant roots where microbial communities work to fix essential nutrients, soil, and improve water absorption in plants in many relationships (5). Microbial communities exist within the rhizosphere, or the soil surrounding the root. This influences a plant's health depending on the type of root system. Fibrous root systems, characteristic of monocots, have numerous thin roots across the upper soil layer emerging from the base of the stem that creates a dense mass of roots. This system primarily prevents soil erosion due to its dense mass roots having a large surface area which allow for water and nutrient absorption (7). As they are disturbed, they are constantly interacting with fresh layers of soil and surface-dwelling microbes, which can cause frequent changes in microbial communities over time with adaptation to water availability, soil composition, and temperature. This interaction can prompt fibrous roots to host more diverse microbial communities. The damage flooding does to these communities has been explored in wheat: a cereal crop with a fibrous root system, showing reduced richness and species composition in bacterial communities (3, 6). The dynamic microbial communities in these root systems may benefit the plant by responding to changes in environmental and soil conditions (7).

Taproot systems, root structures commonly associated with eudicots, have a single thick central root growing downwards from the base of the stem with lateral roots branching off it. The function of this root system is accessing water and nutrients in deeper layers of soil as well as storing nutrients to be used during times of stress such as droughts (8). As they are less disturbed by surface changes due to their depth into the soil, this allows for more stable and consistent microbial communities (7). This can prompt taproot systems to host microorganisms that are adapted to low-oxygen conditions (7). Stable microbial communities can allow for consistent long term plant growth in deep soil for nutrient and water uptake, however, would have a lesser response to rapid environmental and soil conditions (7). Unlike plants with fibrous root systems, the effect that flooding stress has on microbial communities hosted on taproots has not been extensively studied or documented.

In this experiment microorganisms of both fibrous and taproot systems are subjected to flooding stress over a few days to test if there is a significant difference in how these microorganism communities respond to flooding stress. We expect to see a reduction in richness, as well as a change in species composition of microorganisms after both plant species have been subjected to flooding stress over a period of days. Based on the physiology of microbial communities, we hypothesize that there will be a significant decrease in the richness and diversity of the microorganisms of both root systems, however, the magnitude of change will be greater for the fibrous root system as it commonly resides in the hypoxia susceptible layer of the soil.

MATERIALS AND METHODS

To perform this experiment nine *Brassica rapa* and nine *Lolium multiflorum* plants were planted in soil sterilized in a PRO-GROW soil sterilizer for 8 hours at 82.2 degrees Celsius along with a slow-release fertilizer. They were allowed to grow in a temperature and light regulated growth chamber with a light regime of 16 hours light and 8 hours dark at 25 degrees Celsius. The plants were allowed to grow undisturbed outside of water for 16 days so that their root systems could develop.

On day 17, three control samples were taken using the following procedure. The root systems were gently uprooted and surface soil shaken off. The root system was then separated from the above ground structures with flamed scissors and placed in capped conical vials with 10 ml of sterilized distilled water (SDW) and vortexed. Firstly, a streak plate is prepared directly from this solution on Luria broth (LB) agar plates. A 10-5 dilution was performed, and the diluted samples were spread plated on nutrient broth (NB) plates. Nutrient plates were selected as the ideal medium for this experiment as they allow for a non-selective range of bacterial growth whereas Luria broth plates were primarily chosen due to the availability in the lab. Both NB and LB plates are allowed to incubate for two days at 30 degrees Celsius. The number of distinct morphologies is noted and counted to determine a richness factor for each plant's root system and the abundance of each distinct morphology is tracked between plant type, treatment, and replicant.

On Day 17, the other plants are submitted to flooding treatment. After one day (flood1) under flooding three plants are uprooted and NB plates are prepared using the same procedure as the control. On Day 19, since the planting and day 3 (flood3) of the flooding treatment the rest of the plants are uprooted and both NB and LB plates are prepared.

The change in microbial richness and abundance was compared between both plants as well as the different treatment steps: the control, flood one and flood three. This data was analyzed for both alpha-diversity (via a two-way ANOVA) and beta-diversity via NMDS graphs and a non-parametric multivariate analysis of variance using permutations (PERMANOVA) in R (15, 16, 17). The amount of change throughout the flooding treatment and the difference in communities on each plant type allows us to document a difference in response to flooding between the two types of root structures.

RESULTS

To analyze the richness of the samples a two-way ANOVA showed that there was a statistically significant difference between the richness of *L. multiflorum* and *B. rapa* ($p = 0.007181$, Df = 1). The overall richness for both plant types doubled for *B. rapa* and increased by about half for *L. multiflorum* as the flood treatment continued ($p = 0.000263$, Df = 2). This trend can also be seen in the box and whiskers plot (fig. 1). Further organization also showed how the richness for both plant types increased as seen in the overall upward trend in richness as they spend longer under flood conditions, however independently from one another ($p=0.319141$, Df = 2, figure 1). Only on the third day of flooding are there significant similarities between the richness of both organism's microbial communities.

To compare the species composition in the rhizospheres of both plant types before, during, and after exposure to flood stress the results were plotted on a non-metric multidimensional scaling (NMDS) graph and analyzed using a PERMANOVA test. Exploring both taxa and treatment types graphically we can see that there are distinct communities of species in both taxa as well as a change in species composition over the course of the treatment (figure 2). This was also confirmed via a PERMANOVA test which highlighted a statistically significant difference in microorganism composition between both plant taxa ($p= 0.049$, $F = 1.89$, Df=1, $R2 = 0.159$). A PERMANOVA analysis of whether the rhizomal communities differed via treatment without considering the taxa of plant they came from showed that this difference was not statistically significant ($p=0.129$, $f=1.5526$, Df=1,$R2=0.134$).

To further investigate the difference the flooding treatment had on the rhizosphere communities within each plant type an NMDS graph was created for each plant taxa differentiating between taxa (fig. 3, fig. 4). These graphs show in space the change in community caused by the flooding stress. Both species moved along with each axes however, *Brassica rapa* shifted to the right and down along our graph while the ryegrass communities were found in the complete opposite direction after exposure to flooding stress. This showed that the microbial communities of both plants were restructured in vastly different ways throughout the treatment.

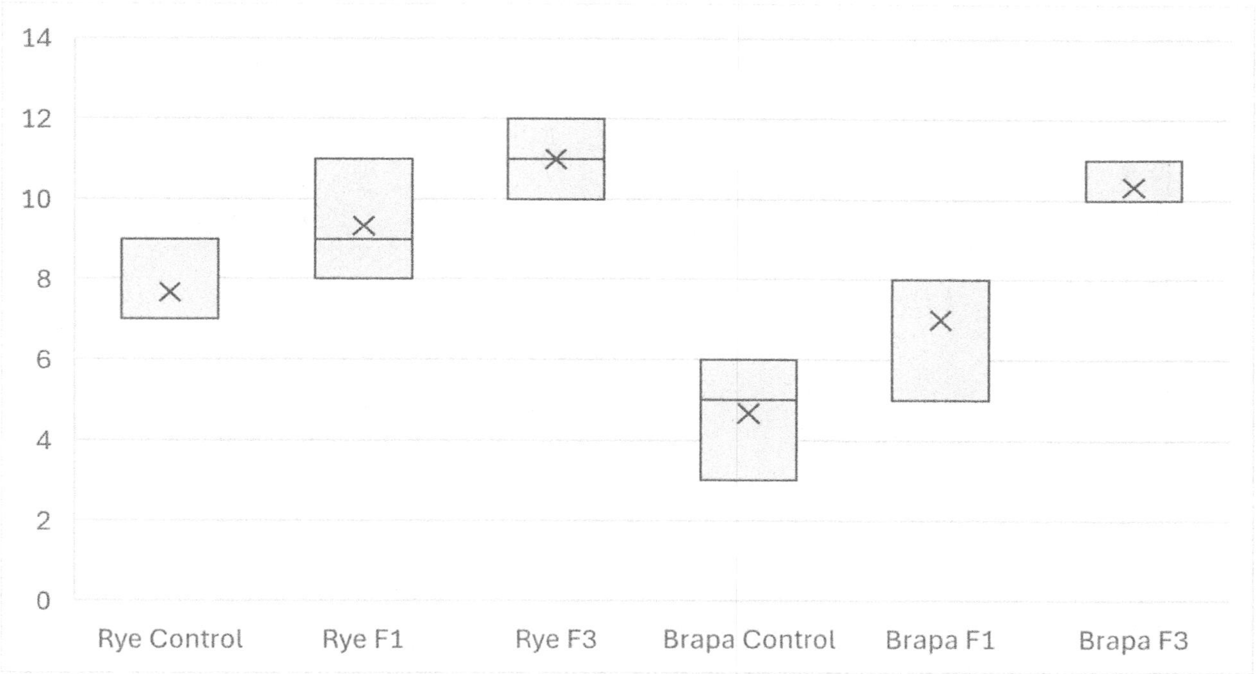

Figure 1. Box and Whiskers plot comparing the microbial richness of spread plates from the roots of *Lolium multiflorum* and *Brassica rapa* at three treatment points. Using letters to show the relationship and significance between samples, the difference between the microbial richness between the two root communities is made clearer

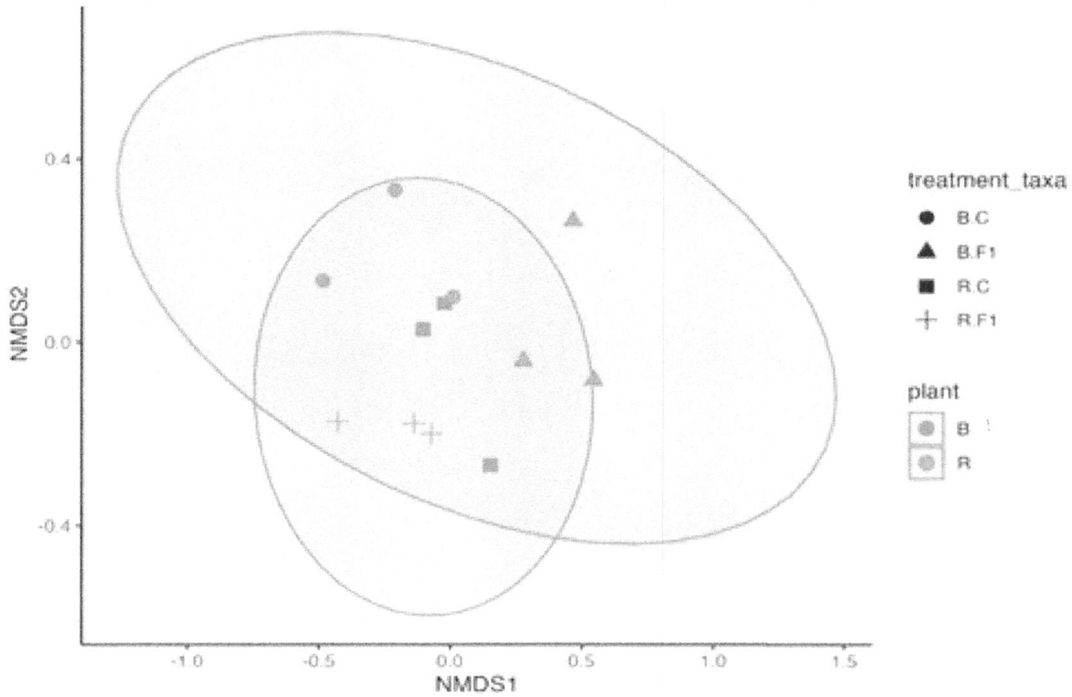

Figure 2. NMDS graph displays the groups microbial communities from each plant type, taxa, at the control and flood 1 treatment points The red ellipse surrounds the data collected from *Brassica rapa* with a 95% confidence which the blue surrounds the *Lolium multiflorum* data with a 95% confidence rate

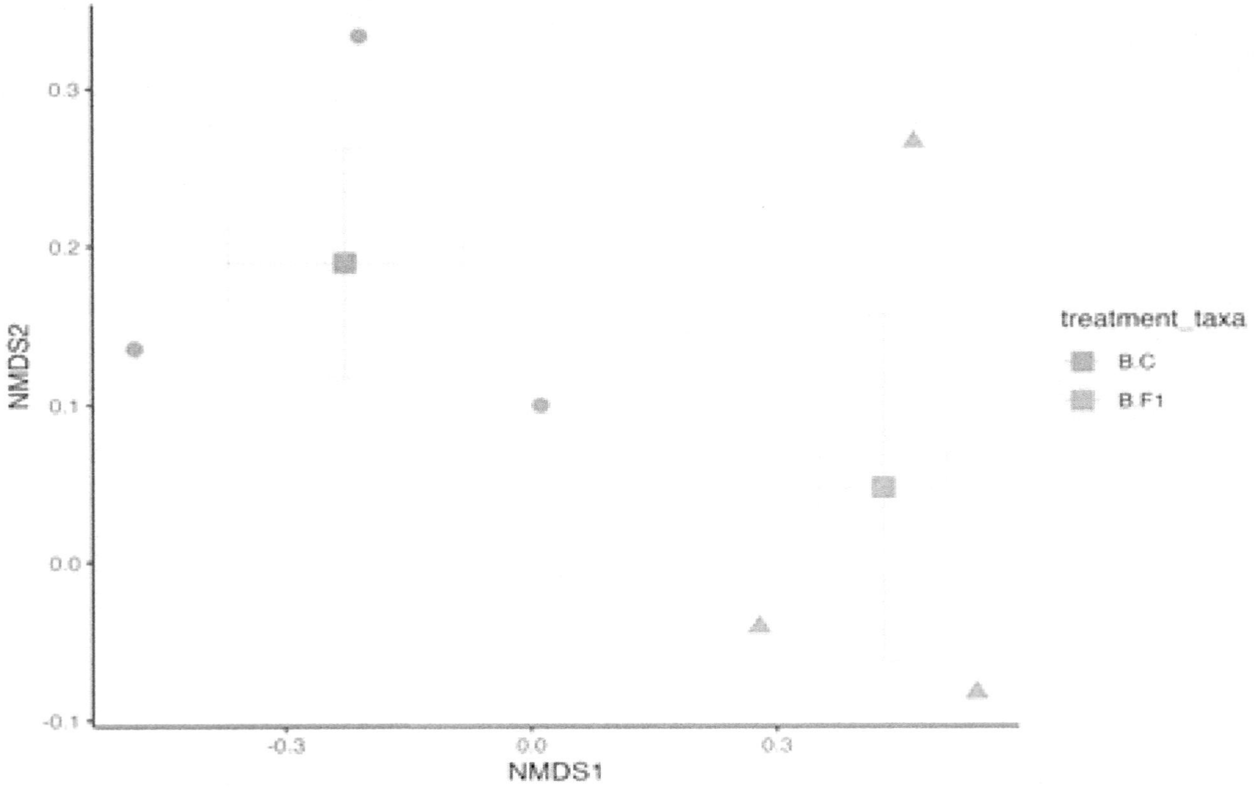

Figure 3. NMDS graph displaying the distinct microbial communities seen in the rhizosphere of *Brassica rapa* before and after flooding treatments. The unflooded, control, treatment are displayed with orange circles, and the results after 24 hours in flooding conditions are shown with blue triangles. For each grouping there is a centroid with error bars showing a 95% confidence rate.

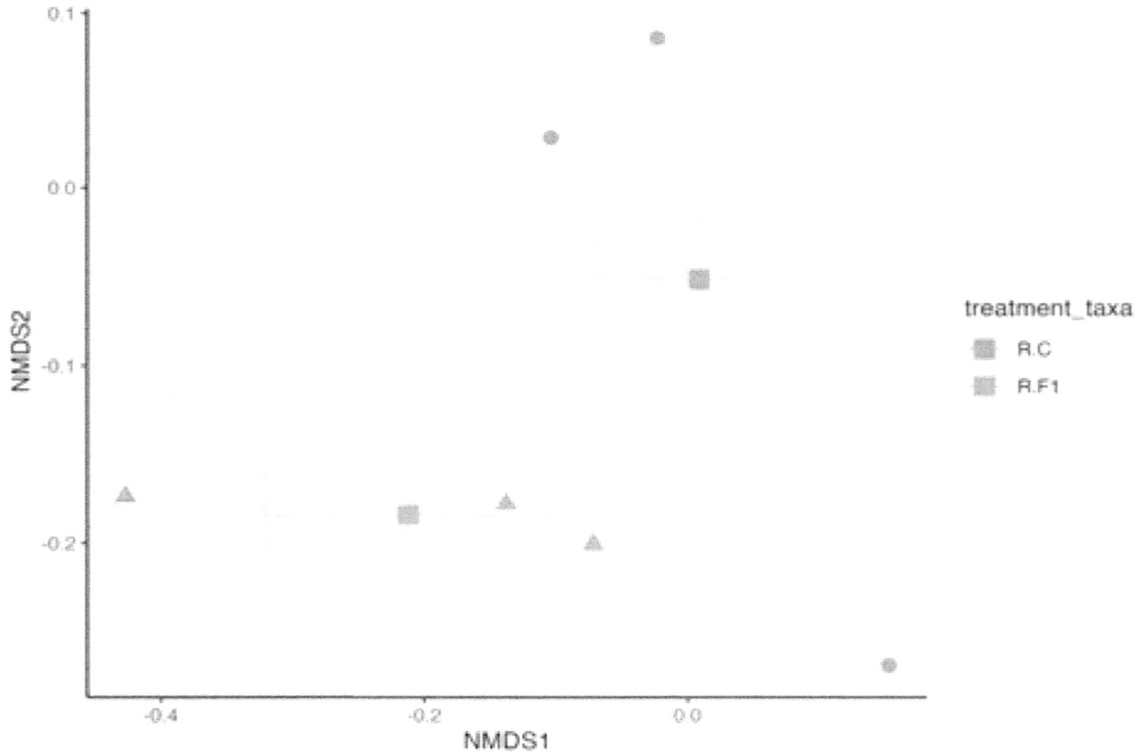

Figure 4. NMDS graph displaying the distinct microbial communities seen in the rhizosphere of *Lolium multiflorum* before and after flooding treatments. The unflooded, control, treatment are displayed with orange circles, and the results after 24 hours in flooding conditions are shown with blue triangles. For each grouping there is a centroid with error bars showing a 95% confidence rate.

DISCUSSION

The microbial communities that inhabit plant roots are one of many crucial factors in plant health and resilience (9,10). As environmental stressors such as flooding become more frequent with climate change, understanding the composition and adaptability of these communities is increasingly important (4). Much of the existing research has focused on the rhizosphere of fibrous root systems, particularly in cereal crops. In our study, we aimed to compare the microbial communities associated with monocotyledonous (fibrous root) and dicotyledonous (taproot) plants, and to investigate how each respond to flooding stress. We hypothesized that there will be a significant decrease in the richness and diversity of the microorganisms of both root systems, however, the magnitude of change will be greater for the fibrous root system.

Our findings revealed distinct microbial communities in the roots of *Lolium multiflorum* (fibrous root) and *Brassica rapa* (taproot), both in terms of richness and species composition. Interestingly, *B. rapa* exhibited a more dramatic shift in microbial community structure under flooding conditions than *L. multiflorum*. This result is contrary to our hypothesis, which predicted that fibrous roots would experience a greater shift. Previous studies on wheat microbiota have shown significant changes in microbial communities due to flooding, which leads to decline in plant health and productivity (6, 11). The shift we observed in *L. multiflorum* aligns with these findings, likely resulting from factors such as changes in soil pH, reduction of oxygen availability, and limited nutrient access in waterlogged soil (6, 12). However, *B. rapa* underwent a more drastic shift in the rhizomal community than *L. multiflorum*, which could be due to differences in root structure or the different communities' individual abilities to withstand flooding stress. In past studies it has been noted that, in parallel to the shift in rhizomal health, there is a direct decline in plant health and productivity in wheat (6, 10). If such a shift affects the health of wheat so significantly, it can be inferred that similar effects would be experienced by plants that have a taproot system (*B. rapa*) which again results showed, underwent a sharper shift in community composition and richness.

A possible explanation is that with fibrous roots being more vulnerable to flooding, *L. multiflorum* may have adapted by establishing a dense root network that helped microbial communities to be stabilized by developing flood-adaptive traits (13). On the other hand, *B. rapa's* taproot being deeper in the soil layer may have created a specialized, less diverse microbial community which disruptions of flooding led to the shift in community structure (14). These structural differences likely influence how each root system moderates microbial composition during and after stress.

A limitation of our study is the short duration of plant growth and flood exposure. With a longer experimental timeline, we might observe more mature rhizosphere development and gain insights into post-flood recovery and resilience.

Future research could expand on this by examining extended flooding periods and assessing how different microbial communities recover after being returned to ideal conditions after an extended period of stress. These findings can help us understand how different root systems shape and respond to microbial community shifts under flooding stress, which is a critical step toward building more resilient agricultural systems in the face of climate change.

LITERATURE CITED

1. Awika J. 2011. Major Cereal Grains Production and Use around the World. Advances in Cereal Science: Implications to Food Processing and Health Promotion. 1:1-13. doi: 10.1021/bk-2011-1089.ch001.
2. Serna-Saldivar S. 2023. Maize. ICC Handbook of 21st Century Cereal Science and Technology. 131-143. https://doi.org/10.1016/B978-0-323-95295-8.00030-7.
3. Francioli D, Cid G, Hajirezaei M-R, Kolb S. 2022. Response of the wheat mycobiota to flooding revealed substantial shifts towards plant pathogens. Frontiers in Plant Science 13.
4. Zhang M, Zhai G, He T, Wu C. 2023. A growing global threat: Long-term trends show cropland exposure to flooding on the rise. Science of the Total Environment. 899. https://doi.org/10.1016/j.scitotenv.2023.165675
5. Bailey-Serres J, Lee SC, Brinton E. 2012. Waterproofing crops: Effective flooding survival strategies. Plant Physiology 160:1698–1709. doi: 10.1104/pp.112.208173
6. Francioli D, Cid G, Kanukollu S, Ulrich A, Hajirezaei M, Kolb S. 2021. Flooding Causes Dramatic Compositional Shifts and Depletion of Putative Beneficial Bacteria on the Spring Wheat Microbiota. Frontiers in Microbiology. 12. https://doi.org/10.3389/fmicb.2021.773116
7. Li J, Yuan X, Ge L, Li Q, Li Z, Wang L, Liu Y. 2020. Rhizosphere effects promote soil aggregate stability and associated organic carbon sequestration in rocky areas of desertification. Agriculture, Ecosystems & Environment 304:107126.
8. Yang H, Zhang Q, Dai Y, Liu Q, Tang J, Bian X, Chen X.

2014. Effects of arbuscular mycorrhizal fungi on plant growth depend on root system: A meta-analysis. Plant and Soil 389:361–374.

9. Thoms D, Liang Y, Haney CH. 2021. Maintaining symbiotic homeostasis: How do plants engage with beneficial microorganisms while at the same time restricting pathogens? Molecular Plant-Microbe Interactions® 34:462–469.
10. Bai B, Liu W, Qiu X, Zhang J, Zhang J, Bai Y. 2022. The root microbiome: Community assembly and its contributions to plant fitness. Journal of Integrative Plant Biology 64:230–243.
11. Martínez-Arias C, Witzell J, Solla A, Martin JA, Rodríguez-Calcerrada J. 2022. Beneficial and pathogenic plant-microbe interactions during flooding stress. Wiley Online Library. 45:2875–2897.
12. Matin N, Jalali M. 2016. The effect of waterlogging on electrochemical properties and soluble nutrients in paddy soils. *Paddy Water Environment.* 15:443-455. https://doi.org/10.1007/s10333-016-0562-y
13. Voesenek, L.A.C.J., & Bailey-Serres, J. (2015). Flood adaptive traits and processes: an overview. New Phytologist, 206 (1), 57-73. https://nph.onlinelibrary.wiley.com/doi/10.1111/nph.13209
14. Fierer, N., Schimel, J.P., & Holden, P.A. (2003). Variations in microbial community composition through two soil depth profiles. Soil Biology and Biochemistry, 35 (1), 167-176. https://www.sciencedirect.com/science/article/pii/S0038071702002511
15. Aslam A, Mahmood A, Ur-Rehman H, Li C, Liang X, Shao J, Negm S, Moustafa M, Aamer M, Hassan MU. 2023. Plant adaptation to flooding stress under changing climate conditions: Ongoing breakthroughs and future challenges. Plants 12:3824.
16. Simmons T, Caddell DF, Deng S, Coleman-Derr D. 2018. Exploring the root microbiome: Extracting bacterial community data from the soil, rhizosphere, and Root endosphere. Journal of Visualized Experiments.
17. Sparks E, Benfey P. 2017.The contribution of root systems to plant nutrient acquisition. Plant Macronutrient use and efficiency. 5:83-92. https://www.sciencedirect.com/topics/agricultural-and-biological-sciences/fibrous-root-system. doi: 10.1016/B978-0-12-811308-0.00005-3.
18. Fierer, N., Schimel, J.P., & Holden, P.A. (2003). Variations in microbial community composition through two soil depth profiles. Soil Biology and Biochemistry, 35 (1), 167-176. https://www.sciencedirect.com/science/article/pii/S0038071702002511

Humboldt Journal of Microbiology
2025, Volume 24

Elucidating the Relationship Between Water Quality and Antibiotic Resistance of Rainwater Microbes Across Western Humboldt County

Theo Murphy, Eve Wendley, Justice Laskowski, Tyler Paredes

ABSTRACT

Limited studies have been conducted regarding the fecal contamination and microbial dispersal through rainwater, with no known studies investigating the potential connection between antibiotic resistant microbes and water quality in rainwater. This study seeks to identify the presence of a correlation between water quality and incidence of antibiotic resistance in rainwater microbes. We hypothesize that there will be a correlation between these two variables. We additionally anticipate a high incidence of fecal contamination across rural and urban collection sites alike, with fecal contamination being more prevalent in rural areas than urban locals. To determine water quality, we measured the concentration of fecal coliforms in four rainwater samples collected throughout western Humboldt County using selective and differential media. We determined the levels of antibiotic resistance using microtiter plates and bacterial growth scoring, quantifying antibiotic resistance by calculating the MIC of each isolate across a concentration gradient of ampicillin. We found no relationship between fecal coliform abundance and the incidence of ampicillin resistance. Each sample contained at least one fecal coliform, denoting fecal contamination of rainwater across all collection sites. Therefore, rainwater should not be consumed or used for domestic use without proper filtration and water treatment. With a high concentration of fecal coliforms in our samples, public health may be compromised, as their presence in rainwater indicates the potential for pathogenic microorganisms to disperse via rain events. Further studies elucidating the role of horizontal gene transfer in rain-dispersed microbes is required to fully understand the impacts rainwater has in the geographic distribution of antibiotic resistant microbes in contaminated water.

INTRODUCTION

Fecal coliforms are indicators of poor water quality that develop in the intestines of humans and warm-blooded animals (1). Even though fecal coliforms do not directly cause disease, they are associated with pathogenic microorganisms. An estimated 1.8 billion people drink from water with fecal contamination all around the world (2). Fecal contamination can be calculated by measuring the concentration of Escherichia coli as well as other members of Enterobacteriaceae (2). Fecal coliform prevalence and concentrations vary between locales. An increase in water contamination occurs in rural areas compared to urban areas and is most common in Africa and South-East Asia's drinking water (2). For example, the Tuojiang river was contaminated due to a variety of sources such as heavy metals, microplastics, and environmental endocrine disruptors (EED), with the primary sources of contamination originating from domestic wastewater and livestock sewage (3). In addition, many known rivers suffer from fecal coliform contamination, indicating poor water quality outside of underdeveloped regions (3).

Due to fecal coliform's prevalence in potable water, concerns for public health have been brought to the forefront of hydrological research. Given that domestic wastewater and livestock sewage are major contributors to the introduction of fecal coliforms into water sources, there are multiple probable sources for fecal coliforms to enter water systems (3). The concerns of poor water quality have been correlated with an increase in stress from local people. Locals have started raising more awareness to the public and government regarding the rising problem of fecal coliforms in their water sources (3). Knowledge is power, and elucidating the role of fecal coliforms in water quality testing allows for the general public to make informed decisions about their health and wellness. In turn, more general knowledge about the impact that the dispersal of contaminated water has on people's health will lead to an increase in research on this topic. Fecal coliforms are not only an issue in contaminated source water but may also be dispersed via rainwater. Most existing works relating to rainwater investigates the safety of roof harvested rainwater (4), inserting a bias in samples with the addition of secondary contamination.

Rainwater quality appears to be dictated by climate and geographic location (5). For example, it has been found that rainwater collected in the wet season is more likely to have higher concentrations of E. coli than rainwater collected in the dry season (5). A study done in Malaysia concluded that it was possible for fecal coliforms to be transmitted through rainwater due to the coliforms they obtained while collecting the roof harvested rainwater (5). This indicates a need for disinfection before consumption (5). Without treatment, such rainwater poses a public health risk if consumed by humans (5). Furthermore, it has been documented that rainwater encounters airborne contaminants and secondary contaminants when in contact with roofs, increasing the concentration of total coliforms detected (5). Establishing a better understanding of rain as a factor contributing to fecal coliform dispersal and associated public health risks is paramount in determining methods to improve human health.

Microbes associated with fecal coliforms can develop resistance to commonly used antibiotics and antimicrobial compounds, potentially leading to an increase in virulence and the need for a more rigorous treatment process (6). Microbes with antibiotic resistance have the capability to share genetic material, either through horizontal or vertical genetic transfer, leading to a greater distribution of antibiotic resistant genes (7), resulting in an increased rate of antibiotic resistance among microbial communities (7). This has led to an estimated 700,000 lives lost annually due to such antibiotic-resistant pathogens (7). While the presence of fecal coliforms in major water sources has been well documented (8), the diversity and antibiotic resistance of microbial communities in rainwater is a major research gap and has not been investigated thoroughly (4). Rainwater transports microbes in the form of precipitation (9), and dispersed microbes may carry genetic material that codes for antibiotic resistance, having the potential to negatively impact public health (3). Should this occur, novel antibiotics would need to be developed to treat the newly-emerging antibiotic resistant bacteria.

This study aims to establish whether there is a correlation between water quality and antibiotic resistance in microbes isolated from rainwater throughout different coastal environments. Despite Humboldt County rainwater being sourced from the Pacific Ocean (10), locality where the rainwater precipitates may be a contributing factor in microbial composition of rain. We hypothesize that there is a correlation between water quality and incidence of antibiotic resistance in rainwater isolates, as fecal coliforms may have a greater potential at spreading antibiotic resistance. We additionally assert that we will find a high concentration of fecal coliforms in rainwater, establishing the need for proper filtration protocols of rainwater intended for personal and domestic use. We investigated fecal coliform abundance and ampicillin resistance in Humboldt County rainwater. Rainwater studies conducted in this geographical area are sparse; documenting the presence of high fecal coliform concentrations and antibiotic resistance in remote areas would highlight the gravity of this public health issue.

Our results serve to document the presence of fecal coliforms and antibiotic resistant bacteria in rainwater. Pathogenic microbes associated with fecal coliforms may infect humans and potentially acquire antibiotic resistance as these pathogens are dispersed, negatively impacting human health. Antibiotics serve as an important defense in combating both minor and life-threatening infections (11). Should our findings lead to a positive indication for fecal coliforms in the presence of microbial communities containing antibiotic resistance genes, public health concerns will arise for residents of western Humboldt County and the greater Pacific West Coast area.

MATERIALS AND METHODS

Monitoring potable water sources is an important step in determining public health and safety. State and federal governments use fecal coliforms as indicators of fecal contamination in freshwater samples. Using fecal coliforms as indicators of water quality spread on Reasoner's 2A agar (R2A) is a conventional method in determining potability of water sources (12). In this study, we followed the standards of determining water quality in drinking water as designated by the United States Environmental Protection Agency (USEPA), where the presence of at least one fecal coliform colony indicates fecal contamination in sites with less than forty samples collected within a one month period (12).

Antibiotic resistance in microbes is not monitored at the government level despite their implications in worsening response time in public health outbreaks (12). As an antibiotic used to target both gram negative and gram positive bacteria alike, ampicillin is a useful baseline for determining antibiotic resistance in bacterial communities (13). Fecal coliforms such as E. coli are gram negative (14), thus generally maintaining resistance to antibiotics specifically targeting the dense peptidoglycan layers of gram positive bacteria (1). Gram negative bacteria are typically affected by beta-lactam antibiotics such as ampicillin due to the antibiotic interfering with cell wall crosslink formation, effectively lysing the cell should the cell attempt to grow (15). Due to ampicillin being used to treat a wide variety of infections in humans, using ampicillin in this study is key to determining how the bacteria will generally react to the environmental introduction of antibiotics.

Collection

To evaluate fecal coliform content in rainwater, we sampled four sites across western Humboldt County during one rain event for a total of four samples. Each rainwater sample was collected using 2 sterile 50mL collection tubes, 2 sterile 65mm funnels, and 1 tube holding apparatus. The apparatus was constructed, holding one tube on each side, and a sterile funnel was placed in each collection tube to maximize rainwater collection. Once fully constructed, the apparatus holds the collection tubes approximately 35 centimeters from the ground. Approximately 1mL of rainwater per site was collected between the two collection tubes. Each sample was preserved at 4°C for 24 hours before processing.

Plating and Bacterial Isolation

In order to quantify the abundance of general culturable microbes and fecal coliforms in our samples, we plated each water sample on R2A and EMB agar respectively. Each sample was vortexed for approximately 5 seconds to achieve a uniform concentration of microorganisms throughout the water. Each rainwater sample was transparent and free of environmental detritus. We spread plate 0.1mL of the undiluted rainwater samples on Reasoner's 2A (R2A) agar and then again on Eosin Methylene Blue (EMB) agar, observing growth after 24 hours and 48 hours of incubation at 37°C. Three representative isolates from each R2A spread plate were chosen and isolated using the streak plate technique on three separate plates. To select the isolates, we designated a number for each colony over the entire plate, randomly selecting three colonies using a number generator. The streak plates were then incubated at 37°C for 24 hours. Observations for each plate were documented, and plates were moved to the refrigerator for short-term storage. Once single colony isolation was achieved and purity confirmed, a single colony of bacteria was transferred to R2A broth and vortexed for a uniform cell concentration across the broth.

Microtiter Broth Dilution

To determine the MIC of each bacterial isolate, we performed a microtiter broth dilution. We dispensed 100µL of Mueller-Hinton broth into each well using a micropipettor. Then, we established a concentration gradient of an ampicillin solution by two-fold dilutions across the microtiter plate, leaving two wells empty for positive and negative controls. The positive control included the Mueller-Hinton broth and the bacterial isolate, excluding ampicillin to allow for bacterial growth. The negative control included Mueller-Hinton broth, excluding both ampicillin and the bacterial isolate to control for sterility (12). We repeated this process eleven more times for a total of twelve isolates tested. The microtiter plates were then incubated at 37°C for 24 hours before scoring the bacteria. All observations were documented after the incubation time, and a MIC was determined for each isolate.

Statistical Analysis

We graphically represented the 24 hour and 48 hour colony counts on R2A and EMB agar plates using bar charts. The layout of the microtiter broth plate was formatted and labeled, with growth ranges and ampicillin concentrations being provided across the rows and columns respectively. We used a table to present a summary of the MICs by site and isolate numbers. The statistical analyses were performed using Google Sheets software, with the p-value being calculated assuming equal variances. The degrees of freedom were calculated using the equation $v=n1+n2-2$, where n1 denotes fecal coliform concentration sample size and n2 represents the MIC sample size.

RESULTS

Bacterial Growth

Each of the four sites had growth on both the Eosin Methylene Blue (EMB) and Reasoner's 2A (R2A) agar plates, indicating growth of coliform and non-coliform bacteria. Colony count was highest on R2A plates in the sample collected from the Humboldt Community Forest, with 478 colonies displayed after a 24-hour incubation period and over 1000 colonies grown after a 48-hour incubation period. The lowest number of colonies cultivated on R2A agar plates was from the sample collected from the Sunny Brae residential area, with 2 visible colonies formed after a 24-hour incubation period and 3 noticeable colonies formed after a 48-hour incubation period. The other sites fall between the two previously mentioned values (Figure 1). On the EMB plates, growth was highest in the agricultural field sample, with 67 colonies after a 24-hour incubation period and 74 colonies after a 48-hour incubation period. The lowest growth was from the Sunny Brae residential area, with 2 colonies formed after 24-hour and 48-hour incubation periods. The other two sites displayed colony growth between the two previously stated ranges (Figure 2). Only the EMB plate from the Sunny Brae residential area showed no additional growth after the 24 hour mark. Morphologies of colonies were extremely variable in shape, color, and associated metabolic function (Appendix 1).

Antibiotic Resistance

The MIC was determined using two-fold dilutions of ampicillin across ten microtiter wells, starting at a concentration of 200 µg/mL and ending with a concentration of 0.78 µg/mL (Figure 3). To account for possible microbial contamination of the microtiter plate, a negative control was established using solely Muller-Hinton broth. Utilizing Mueller-Hinton broth and bacterial isolates, a positive control was established for each concentration gradient. Ampicillin resistance and susceptibility w. The threshold for ampicillin resistance was placed at 100 µg/mL (1x the working concentration of antibiotics). In instances where no bacterial growth occurred, the bacterial isolate was denoted as being susceptible to ampicillin. However, there were two cases of contamination with the negative control reporting bacterial growth (Figure 3).

To determine if there was a relationship between the MIC (µg/mL) and fecal coliform concentration (cells/mL), comparisons were made utilizing a 2-tailed t-test with the assumption of equal variance. The two instances of bacterial growth in the negative control well were excluded from statistical analysis (Figure 4). The resulting analysis was not statistically significant (t=0.062).

Two bacterial isolates were found to exhibit ampicillin resistance; one ampicillin-resistant isolate was collected from Samoa Beach while the other ampicillin-resistant isolate was collected from the Sunny Brae residential area. Three bacterial isolates were completely susceptible to ampicillin, with one ampicillin-susceptible isolate being collected from Samoa Beach, while the other two ampicillin-susceptible isolates were collected from agricultural land. Every other CFU isolated across sites exhibited a range of bacterial growth falling between the two extremes of resistance and susceptibility (Table 1).

Site 1 is Community Forest (478 colonies at 24 hours and over 1000 at 48 hours), site 2 is Samoa Beach (16 colonies at 24 hours and 101 at 48 hours), site 3 is Sunny Brae residential area (2 colonies at 24 hours and 3 at 48 hours), and site 4 is an agricultural field (197 colonies at 24 hours and 213 at 48 hours).

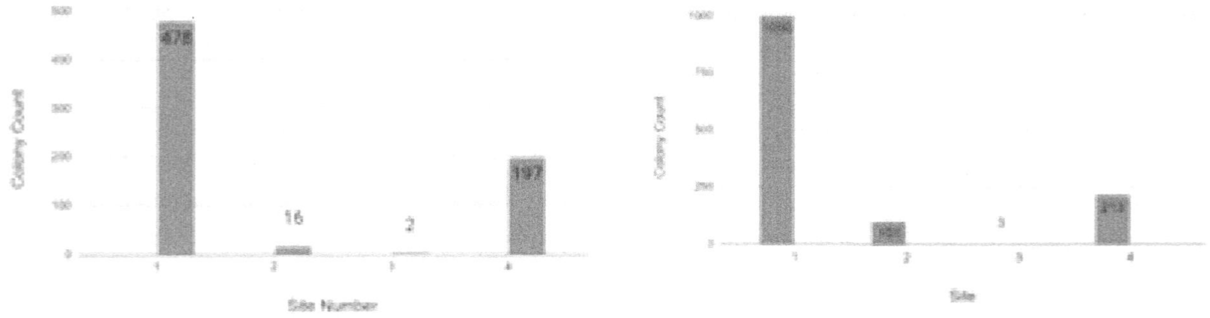

Figure 1. Colonies Formed on R2A Agar after 24 and 48 Hours Incubation at 37°C.

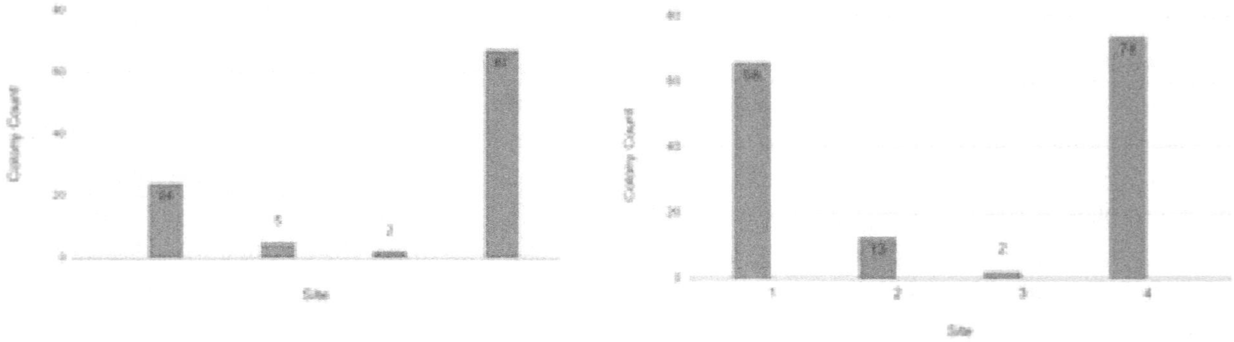

Figure 2. Colonies Formed on EMB Agar after 24 and 48 Hours Incubation at 37°C. Site 1 is the Humboldt Community Forest (24 colonies at 24 hours and 66 at 48 hours), Site 2 is Samoa Beach (5 colonies at 24 hours and 13 at 48 hours), Site 3 is Sunny Brae residential area (2 colonies at 24 hours and 2 at 48 hours), and Site 4 is an agricultural field (67 colonies at 24 hours and 74 at 48 hours).

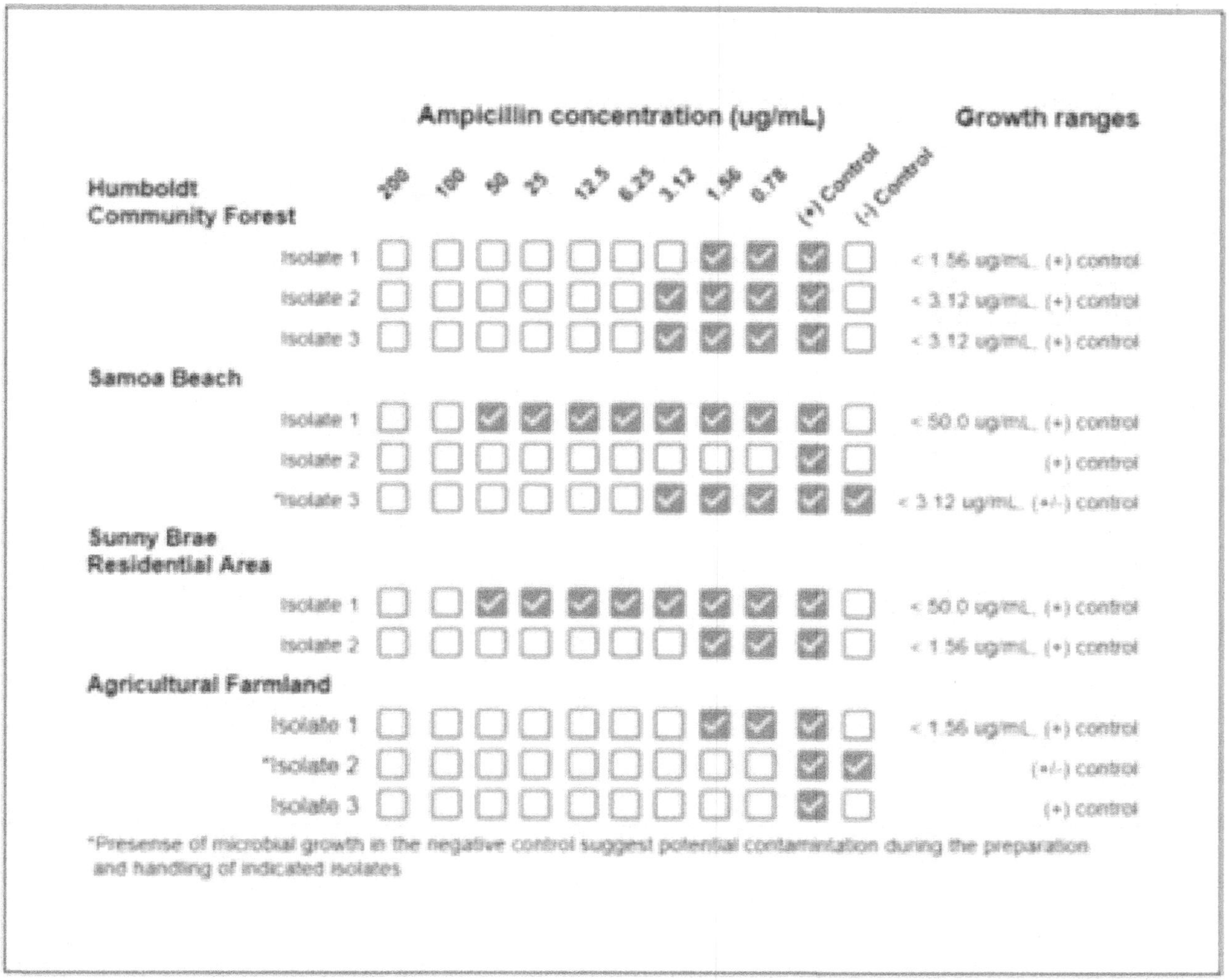

Figure 3. Results of MIC on Bacterial Isolates. Isolate tests conducted on a 96-well plate with combination of ampicillin, Mueller-Hinton broth, and bacterial isolate. Positive control lacked ampicillin and negative control lacked bacterial isolate. Two instances of contamination occurred in Isolate 3 at Samoa Beach and Isolate 2 at the Agricultural Farmland.

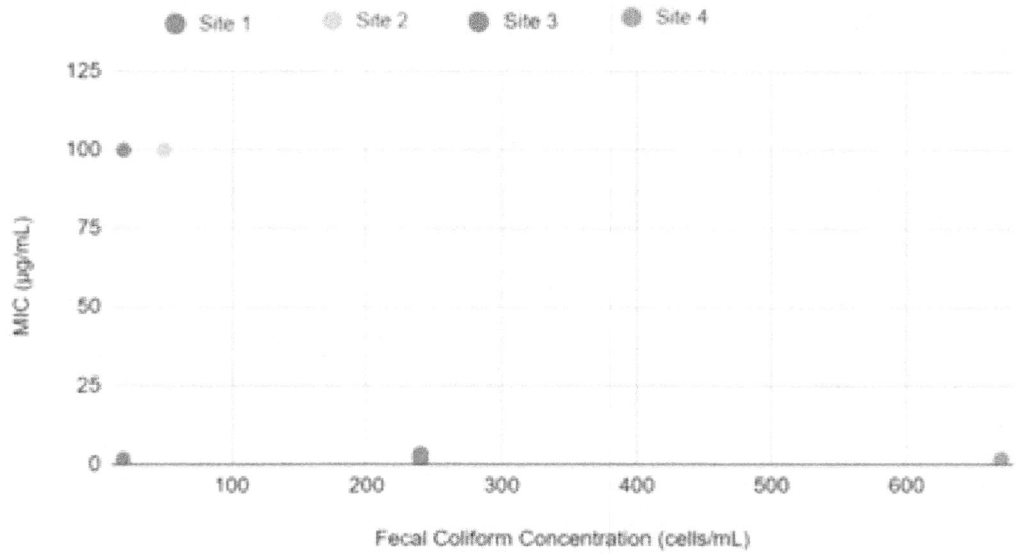

Figure 4. MIC (µg/mL) vs Fecal Coliform Concentration (cells/mL). Isolates were colored by site. The 2-tailed t-test found no correlation between fecal coliform concentration and MIC ($t=0.062$), with equal variances being assumed ($\nu=14$).

Table 1 Minimum Inhibitory Concentration (MIC) of coliforms isolated from rainwater. (R) means Resistant, (S) means Susceptible.

Origin of Bacteria	Antimicrobial Agent	MIC (μg/mL)
Humboldt Community Forest 40.87077 N, 124.07468 W Isolate 1	Ampicillin	1.5625
Humboldt Community Forest 40.87077 N, 124.07468 W Isolate 2	Ampicillin	3.125
Humboldt Community Forest 40.87077 N, 124.07468 W Isolate 3	Ampicillin	3.125
Samoa Beach 40.821641 N, -124.1899812 W Isolate 1	Ampicillin	100 (R)
Samoa Beach 40.821641 N, -124.1899812 W Isolate 2	Ampicillin	No growth (S)
Samoa Beach 40.821641 N, -124.1899812 W Isolate 3	Ampicillin	3.125
Sunny Brae Residential Neighborhood 40.86936 N, 124.06957 W Isolate 1	Ampicillin	100 (R)
Sunny Brae Residential Neighborhood 40.86936 N, 124.06957 W Isolate 2	Ampicillin	1.5625
Agricultural Farmland 40.86938 N, 124.06961 W Isolate 1	Ampicillin	1.5625
Agricultural Farmland 40.86938 N, 124.06961 W Isolate 2	Ampicillin	No growth (S)
Agricultural Farmland 40.86938 N, 124.06961 W Isolate 3	Ampicillin	No growth (S)

DISCUSSION

Every year, nearly 5.7% of diseases and 4% of deaths worldwide are attributed to lack of clean drinking water (4). Therefore, assessing water quality in prospective domestic water sources is imperative to keeping rural human communities healthy. Antibiotic resistance may complicate cases of infection as a result of water contamination. Genes coding for antibiotic resistance may be transferred via horizontal gene transfer among microbes within a community. The genetic information packed within chromosomes and plasmids can be readily spliced via transformation, transduction, and conjugation from an adjacent bacterial cell with its own distinct genetic composition (16), a mechanism conserved in bacteria to adapt to changing environments. Should fecal coliforms be discovered in rainwater isolates consisting of antibiotic-resistant microorganisms, coliforms would pose as a potential vector responsible for the integration of such genetic material into resistant microbes.

Cultured in R2A agar for 24 hours, the first major finding revealed colony forming units (CFUs) of different morphologies, ultimately confirming the presence of oligotrophic bacteria existing in rain. Using EMB plates, the residential site denoted the highest water purity out of all of our sampling sites, with only two fecal coliform colonies forming after 48 hours of incubation, equating to approximately 20 cells/mL in the original rainwater sample. Improved water quality is often associated with densely populated areas (2), resulting in a lower concentration of fecal coliforms in this sample. Conversely, our rural sampling sites displayed elevated fecal coliform concentrations as is congruent with our hypothesis. EMB agar plates exhibited morphological diversity, indicating varying degrees of lactose fermentation, a key characteristic of coliform metabolic activity (7). Lastly, our research found fluctuating patterns of antibiotic resistance throughout the collection sites and among bacterial communities existing within the same site.

The scatterplot (Figure 4) demonstrates an inverse relationship, where greater antibiotic resistance was detected in rain isolated from environments with lower fecal coliform concentrations. Furthermore, we found little evidence supporting a relationship between the two variables with a statistically insignificant correlation ($p>0.05$). Therefore, a meaningful linear relationship cannot be established with our present data. In light of a relatively small data set, this weak correlation coefficient is likely attributed to the site location where rainwater was isolated. In agricultural farmland or rural areas where human populations are relatively minimal, we see high coliform concentration imputable to wildlife abundance and lack of well-established waste removal systems. In contrast, low coliform concentrations are reflected in urbanized locations presumably due to the presence and activity of waste removal systems and water treatment plants.

The MIC values (Table 1) correspond to greater patterns of antibiotic resistance in bacterial isolates collected from heavily trafficked and/or densely populated locations. This is noteworthy, in that the data suggests ampicillin resistance is not directly associated with coliform concentration and water purity, but instead potentially indicates the presence of diverse microbial communities within these particular collection sites.

This research does not support our main hypothesis, as we found no evidence of direct correlation between fecal coliform concentration and incidence of ampicillin-resistant bacteria in rainwater. However, we found evidence of high concentrations of fecal coliforms, with as many as 74 distinct colonies of fecal coliforms forming on the agricultural farmland's EMB plate after a 48-hour incubation period; this colony count translates to 740 cells/mL within our sample, demonstrating severe fecal contamination of rainwater in rural areas in western Humboldt County. In accordance with our secondary hypothesis, we found evidence of fecal contamination of rainwater across all sites.

Incidence of ampicillin-resistant bacteria greatly varied between individual microbes, with no detectable trend between site and MIC. Due to differences in MIC between individual rainwater isolates, we cannot accurately ascertain the role horizontal gene transfer plays in the transmission of antibiotic resistance among rain-dispersed bacteria. While we assert that precipitation in the form of rain may bind and disperse antibiotic resistant microbes via coastal drift (17), the transfer of genetic material coding for ampicillin resistance between rain-dispersed bacteria at the community level remains elusive. Variance in MIC may be contributed to the slow growth characteristic of oligotrophic bacteria. Additionally, having less genetic material may be advantageous in microbes that frequently inhabit low-nutrient environments, such as the Pacific Ocean where Humboldt County rain originates (10). Introducing genetic information encoding for antibiotic resistance may prove costly to such oligotrophic microorganisms who need to conserve energy for survival. Furthermore, specific MICs associated with varying levels of antibiotic resistance may be taxon-specific, as we cannot be certain that organisms with similar MICs did not belong to the same species or genera. An analysis of bacterial community composition associated with ampicillin-resistant microbes in rainwater may further enhance our findings.

Many recently-emerging studies have revealed that the rapid rise of antibiotic resistance has been a primary result of misuse and over-use of antibiotic and antimicrobial compounds in medicine and agriculture (4). For instance, researchers in

Poland investigating multi-drug resistant bacteria suggest that through the course of evolution, bacteria has developed a variety of mechanisms counteracting the effects of antibacterial agents (6). A separate review exploring drug resistance in bacteria isolated from a multi-profile hospital further supports these claims (15). Consequently, this resistance has led to reduced efficacy of once reliable medications, seen in an increase in cases of untreatable bacterial infections in humans. Coastal water quality monitoring programs typically detect a narrow spectrum of bioindicators, consequently overlooking antibiotic resistant microbes employing rain as a vehicle to enter natural waterways and potable water sources (18). Investing research with a primary focus on specific vectors and dispersal methods of antibiotic resistant microbes is critical to the preservation of human health and wellbeing; such research allows the public to make informed decisions regarding their personal and domestic water use.

Our research was designed to eliminate secondary sources of contamination, leading to improved data quality and accurate counts of ampicillin resistant bacteria isolated from rain. Eliminating contamination biases among our samples directly targeted a major research gap in current literature. Microbial abundance and composition may be site-dependent, adding a limitation to our study. The choice of ampicillin as the sole antibiotic used in MIC testing offers additional limitations in ascertaining overall antibiotic resistance in rain-dispersed bacteria. Further limitations include collection period and associated constraints, where the sample size was smaller than desired due to a narrow collection period and unpredictable weather patterns.

To develop strategies that combat the global crisis of antibiotic resistance, future work may further investigate other relative factors influencing the movement and pathways of such oppositional microorganisms such as anthropogenic waste streams throughout urbanized regions. Waste streams are hot spots for accumulations of microbial diversity and abundance, as they transport industrial, commercial, and hazardous waste (18) sourced from random locations that inhabit various microbial species. Densely populated regions consist of laboratories, hospitals, and other establishments where antimicrobial compounds are routinely dispensed. Upon elimination and provided with appropriate weather conditions such as rain, such compounds may use urine or feces as a vehicle to enter waste streams. Considering the vast amounts of contamination and bacterial growth in urbanized waste streams, it would be beneficial to investigate and identify specific vectors and environmental parameters that enable antibiotic-resistant microbes to develop, thrive, and disperse.

LITERATURE CITED

1. You S, Huang X, Xing L, Lesperance M, LeBlanc C, Moccia L, Mercier V, Shao X, Pan Y, Zhang X. 2023. Dynamics of fecal coliform bacteria along Canada's coast. Marine Pollution Bulletin. 189:114712
2. Bain R, Cronk R, Hossain R, Bonjour S, Onda K, Wright J, Yang H, Slaymaker T, Hunter P, Prüss-Ustün A, et al. 2014. Global assessment of exposure to faecal contamination through drinking water based on a systematic review. Trop Med Int Health. 19(8):917–927. doi:10.1111/tmi.12334.
3. Xu G, Wang T, Wei Y, Zhang Y, Chen J. 2022. Fecal coliform distribution and health risk assessment in surface water in an urban-intensive catchment. Journal of Hydrology 604:127204
4. Anindita KM, Rahman A, Alim M, Xiong C, Hossain S, Sathasivan A. 2025. Microbial ecology of harvested rainwater: Assessing quality, antimicrobial resistance and geographical variation. Journal of Cleaner Production 486:144439
5. Leong JYC, Chong MN, Poh PE, Hermawan A, Talei A. 2017. Longitudinal assessment of rainwater quality under tropical climatic conditions in enabling effective rainwater harvesting and reuse schemes. Journal of Cleaner Production. 143:64–75. doi:10.1016/j.jclepro.2016.12.149.
6. Urban-Chmiel R, Marek A, Stępień-Pyśniak D, Wieczorek K, Dec M, Nowaczek A, Osek J. 2022. Antibiotic Resistance in Bacteria. Antibiotics 11(8):1079. doi.org/10.3390/antibiotics11081079
7. Wen L, Cui Y, Huang L, Wei C, Wang G, Zhang J, Jiang Y, Wei Y, Shen P. 2024. Changes of composition and antibiotic resistance of fecal coliform bacteria in municipal wastewater treatment plant. Journal of Environmental Sciences. 146:241–250. doi:10.1016/j.jes.2023.09.012.
8. Grabowski AT, Jozwiakowski K, Bochniak A, Stachyra P, Radlinski B. 2023. Assessment of Rainwater Quality Regarding Its Use in the Roztocze National Park. Applied Sciences 13(10):6110. doi.org/10.3390/app131061-10
9. Bouchali TR, Mandon C, Danty-Berger E, Geloen FA, Marjolet L, Youenou B, Pozzi AC, Vareilles S, Galia W, Kouyi GL, Toussaint JY, Cournoyer B. 2024. Runoff microbiome quality assessment of a city center rainwater harvesting zone shows a differentiation of pathogen loads according to human mobility patterns.

International Journal of Hygiene and Environmental Health 260:114391

10. Beaudin E, Lorenzo ED, Miller AJ, Seo H, Joh Y. 2023. Impact of Extratropical Northeast Pacific SST on U.S. West Coast Precipitation. Geophysical Research Letters 50(3):doi.org/10.1029/2022GL102354
11. Nappier SP, Liguori K, Ichida AM, Stewart JR, Jones KR. 2020. Antibiotic Resistance in Recreational Waters: State of the Science. Int J Environ Res Public Health 17(21):8034. doi: 10.3390/ijerph17218034
12. Hancock, R. E. W. 1999, Hancock Laboratory Methods. Department of Microbiology and Immunology, University of British Columbia, British Columbia, Canada. MIC Determination By Microtitre Broth Dilution Method. https://cmdr.ubc.ca/bobh/method/mic-determination-by- microtitre-broth-dilution-method/
13. Zhao R, Feng J, Huang J, Li X, Li B. 2021. Responses of microbial community and antibiotic resistance genes to the selection pressures of ampicillin, cephalexin and chloramphenicol in activated sludge reactors. Science of The Total Environment 755:142632
14. Luca NV, Carginale V, Supuran C, Capasso HC. 2022. The gram-negative bacterium Escherichia coli as a model for testing the effect of carbonic anhydrase inhibition on bacterial growth. Journal of Enzyme Inhibition and Medicinal Chemistry 37:2092-2098
15. Maczynska AB, Frej-Madrzak M, Sarowska J, Woronowicz K, Choroszy-Krol I, Jama-Kmiecik A. 2023. Evolution of Antibiotic Resistance in Escherichia coli and Klebsiella pneumoniae Clinical Isolates in a Multi-Profile Hospital over 5 Years (2017–2021). JCM 12(6):2414. doi.org/10.3390/jcm12062414
16. Madigan MT, Bender KS, Buckley DH, Sattley WM, Stahl DA. 2018. Brock Biology of Microorganisms, P. 316-320. 15th ed, Pearson, New York, Ny.
17. Denissen JK, Reyneke B, Waso M, Khan S, Khan W. 2021. Human Pathogenic Bacteria Detected in Rainwater: Risk Assessment and Correlation to Microbial Source Tracking Markers and Traditional Indicators. Frontiers in Microbiology 12:659784
18. Carney RL, Labbate M, Siboni N, Tagg KA, Mitrovic SM, Seymour JR. 2019. Urban Beaches Are Environmental Hotspots for Antibiotic Resistance Following Rainfall. Water Research 167: 115081 https://doi.org/10.1016/j.watres.2019.115081
19. Lal, A., & Cheeptham, N. 2007, Eosin-Methylene Blue (EMB) Agar. American Society for Microbiology. https://asm.org/protocols/eosin-methylene-blue-agar-plates-protocol

Appendix 1 (Colony Morphology of Bacteria)

Colonies on R2A media possessed various morphologies. They ranged in color from yellow to beige to white and margins from circular, convex, and irregular. On EMB plates the colonies additionally demonstrated various morphologies. Colors ranged from white, pink, red, blue, and purple. Black and purple colonies are coliform species that rapidly ferment lactose, with further tests necessary to determine genus and species, while the red and pink colonies do not ferment lactose (19).

Humboldt Journal of
Microbiology
2025, Volume 24

Use of *Enterococcus faecalis* and Cetylpyridinium Chloride in Altering Canine Mouth Microbiome

Curtis Cline, Quigley Espinola, Sofia Lopez-Orenday, Scarlet Renner, Alex Yasumatsu

ABSTRACT

Like most mammals, canines have copious amounts of bacteria in their oral cavity. While some of these bacteria are beneficial and even necessary to their health, overgrowth of harmful bacterial strains can lead to plaque buildup and gum disease. To combat plaque and bacteria, several products are marketed towards canines. In this study, we aim to explore how the microbiome of the canine mouth is affected by a potential probiotic, *Enterococcus faecalis*, and an antimicrobial water additive found in dog mouthwash products, Cetylpyridinium chloride monohydrate (CCM). We hypothesized that both would be successful in their ability to decrease harmful bacteria. Their relative effectiveness is lacking in literature, however, we expect the CCM to be more effective due to its antimicrobial properties. In this study, we explore this idea by isolating what we believe to be three strains of environmental bacteria based on morphology from a canine's mouth and assessing their ability to grow in nutrient broth containing the probiotic and CCM. We then plated these samples and calculated the original concentration of bacterial growth. We found a significant difference in the abundance of bacteria between the two treatments and the control. CCM inhibited all growth, and *Enterococcus faecalis* significantly decreased bacterial growth. Both the CCM and *Enterococcus faecalis* show promising results on adjusting the mouth's microbiome. CCM could potentially harm gut bacteria, making its use in water additives for dogs potentially harmful as it would reach their digestive systems.

INTRODUCTION

The mammalian oral microbiome is a complex community of microorganisms that play a critical role in maintaining oral health and contributing to overall health. Oral microbiota impact oral health through periodontal diseases, which are gum diseases such as gingivitis, due to colonization of harmful bacteria (1). Additionally, microbiota can lead to dental caries and other endodontic, or inner tooth, infections through acidogenic and acidophilic bacteria (2). Oral health can also influence the health of organs in the body. The microbiome of the gut is directly associated with oral microbiota. This impacts how food is digested as well as gastric infections that can develop due to pathogens in dental plaques (3). Research has shown there is a link between poor oral health to oral and non-oral cancers (4) and Rheumatoid arthritis (5). Despite the growing knowledge of the human oral microbiome and its subsequent systematic impacts, research into the oral microbiome of canines has been relatively limited. Further study can provide insight into shared disease mechanics and microbial transmission between humans and pets. As well as filling gaps in our understanding of the bodily impacts of oral microbes in canines.

There is a large variety of bacteria found in the canine oral microbiome. Of those that can be isolated, there are approximately 37 genera of bacteria (6). The most common bacteria found in healthy canines include clades of aerobic species of *Actinomyces, Porphyromonas*, and *Campylobacter* (7). The diversity of the oral microbiome is influenced by a variety of factors such as diet and the inclusion of probiotic supplements (8, 9), age (10), and dental care through dental chews or mouth rinses (11, 12). Overall, the oral microbiome plays a crucial role in both oral and systemic health, highlighting the importance of maintaining oral hygiene for overall well-being.

Due to the crucial nature of sustaining a healthy oral microbiome, it's important to continue to develop our understanding of the microbiota that reside there. It's beneficial to know how we might be able to better regulate harmful bacteria that cause periodontal disease (1, 2) while still being able to facilitate a habitable environment for the other bacteria that reside there. Applying research to canine health practices and products allows researchers to further develop evidence-based solutions that improve pet health and prevent future diseases (11, 12). Providing information about oral microbiota structure and function allows companies to better develop products that effectively combat them (13). It also allows veterinarians to be more knowledgeable in what treatments they recommend and gives a wider variety of options for treatment plans. With this knowledge, we can potentially improve the quality of life and possibly the length of life for our pets (3, 4, 5).

There are several methods used to attempt to control the microbiome within dogs. One of the methods is through the use of probiotics, which work to strengthen the communities of beneficial bacteria within the mouth microbiome. The benefit of a probiotic is that it does not kill beneficial bacteria; instead strengthens them, making it harder for the pathogenic bacteria to grow. It does not directly kill the pathogenic bacteria, but rather is outcompeted by the probiotic bacteria (14). The other commonly used strategy to promote a healthy mouth is through the use of antimicrobials, which contribute to the death of bacteria. Unlike the probiotic, antimicrobials do kill bacteria, usually with little distinction between beneficial and harmful bacteria in the microbiome. The major drawback of an antimicrobial, however, is that after the antimicrobial is taken away, it is easier for the microbiome of the mouth to change, since there are fewer established bacterial colonies present, which can lead to infection, and if the gut microbiome is compromised, even sepsis (15).

In this study, we look into a possible probiotic, a bacterium called *Enterococcus faecalis*, which is closely related to a commonly used probiotic called *Enterococcus faecium* (9,14). Probiotics are normally given as an added supplement to canine food. Enterococci are the largest genus of lactic acid bacteria, which are natural inhabitants in the gut flora of animals and humans. Probiotics such as those in the Enterococci genus have been shown to improve protective immune responses and inhibit enteropathogens such as *Escherichia coli* (16). *E. faecium* is more popularly marketed as a probiotic for canines, however, one company, SYMBIOFLOR 1, does use *E. faecalis* instead. The mechanics of how Enterococci bacteria benefit the body are still being studied. It is believed that the transposon- or plasmid-mediated gene transfer ability of Enterococci is what allows them to bind and connect with other cells and create protective barriers (17). This is in addition to Enterococci outcompeting pathogens that can appear in a microbiome.

In addition to *Enterococcus faecalis*, we will be looking at CCM. This chemical is a known antimicrobial found in a variety of oral health products for both humans and animals. Previous research has found that CCM can prevent or slow the growth of oral plaque. Specifically, it can inhibit the growth of gram-negative obligate anaerobic microorganisms embedded in deep dental biofilms (11). It has been suggested that CCM can exert anti-bacterial effects in oral biofilms by diffusion through the biofilms. Additionally, the local connectivity, size, and microbiota-wide connectivity of bacterial networks can be reduced by a CCM treatment.

For this experiment, we intend to use nutrient agar plates and nutrient broth tubes to culture canine oral bacteria samples, as opposed to using canine models for treatment. We will be treating canine oral bacterial cultures with *E. faecalis* and CCM. Both treatments will be useful in comparing the growth of bacteria in a controlled environment simulating the body temperature of canines. Thus providing evidence of the effectiveness of bacterial control attributed to each of the treatments.

Here we are testing the effectiveness of CCM and *Enterococcus faecalis* in controlling microbial growth. If CCM is an effective antimicrobial, it will drastically reduce the presence of bacteria in an agar plate. If probiotic bacteria are an effective treatment for oral microbiomes, then they will outcompete isolated environmental bacteria and cause them to decrease in abundance. We expect there to be at least a 50% decrease in the number of colony-forming units in the treated spread plates compared to the control. We will be taking oral samples from a canine and isolating three unique bacteria, and then applying our treatments to see their effect. The information gathered from this experiment will be valuable in the education of pet owners on managing the oral health of their pets.

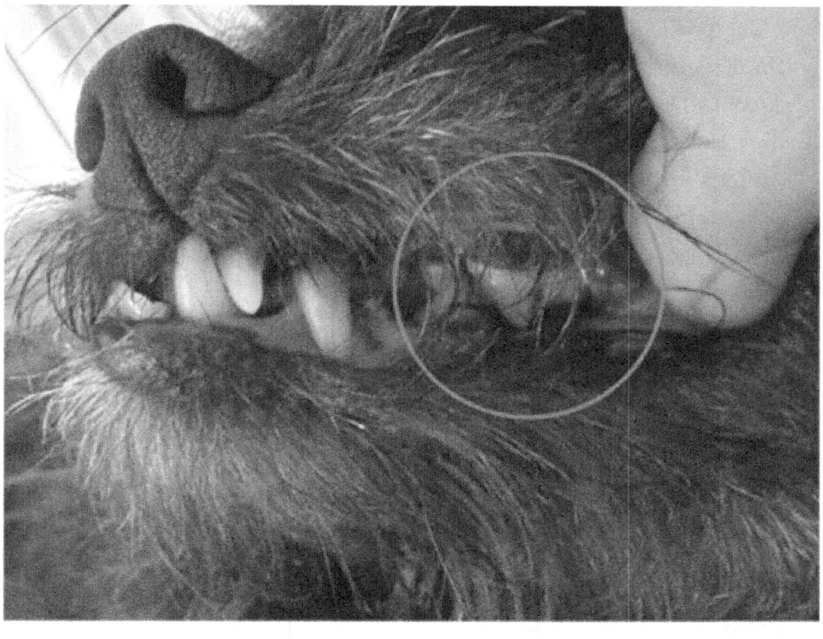

Figure 1. Our sample was collected from the upper carnassial tooth (circled in red) from Axel (*Canis familiaris*), an 8 year old 14 lbs male terrier mix

MATERIALS AND METHODS

We determined that an in vitro model would be best in order to isolate single strains of bacteria for treatment, rather than looking at bacterial community interactions in an in vivo model. This way, we can determine which strains are more affected by antimicrobial or probiotic treatment (18). We can also control the conditions in which the bacteria are cultured and reduce variability found in animal models, such as differences between the hosts' immune systems and oral environments (18). In vitro models eliminate the ethical issues that come with using animal models. This will allow us to easily replicate the experiment and infer the causation of the results. Since we do not yet know the full effects of antimicrobial and probiotic treatments on canine health, we can prevent the risk of harm by replacing an animal model with an in vitro alternative. We also eliminate the need for training on the appropriate handling of animals for experimental testing (19).

We used a sterile swab to obtain our canine oral sample from the upper carnassial tooth of an 8 year old terrier mix (Fig.1). The sample was inoculated into nutrient broth and incubated at 37°C for 48 hours. The broth culture was used to create a TSA streak plate, which was incubated at 37°C for 48 hours. Using a sterile loop, three colonies with different morphologies were isolated from the streak plate. Each colony was used to make LB streak plates, which were incubated at 37°C for 48 hours. Plates were observed to determine if each bacterial strain was unique based on its morphology. This process was repeated from the original broth culture of the canine oral sample until three morphologically unique strains were obtained (Fig. 2).

Our probiotic bacteria culture, *Enterococcus faecalis*, was obtained from the Cal Poly Humboldt microbiology stockroom. This culture was transferred to fresh nutrient broth throughout the experiment to maintain viability.

After three unique strains were obtained, each strain was transferred from the plate to a nutrient broth using a sterile loop. After incubation, the broths were used to create three replicate nutrient broths for each strain. This was repeated to make a total of six nutrient broths for each strain, three of which were treated with the antimicrobial and three that were treated with the probiotic. We also created a control nutrient broth of each strain, which would not receive a treatment.

For our antimicrobial treatment, we used CCM. Following the concentration seen in commercial canine mouthwash, we weighed 4.6 mg of CCM to be added to 5 ml of broth for each strain. CCM was added prior to the inoculation of the bacterial strain into the nutrient broth replicates.

For our probiotic treatment, nutrient broth replicates were inoculated with *Enterococcus faecalis* using a sterile loop prior to the inoculation of each bacterial strain. The control broths and broth replicates that were treated with the antimicrobial and the probiotic were incubated at 37°C for 48 hours.

Following incubation, we took a broth of each strain and treatment (Strain 1/CCM, Strain 2/CCM, Strain 3/CMM, Strain 1/Ef, Strain 2/Ef, Strain 3/Ef) as well as each control broth (Strain 1, Strain 2, and Strain 3) to create serial dilutions for spread plating. Using a spectrophotometer to measure

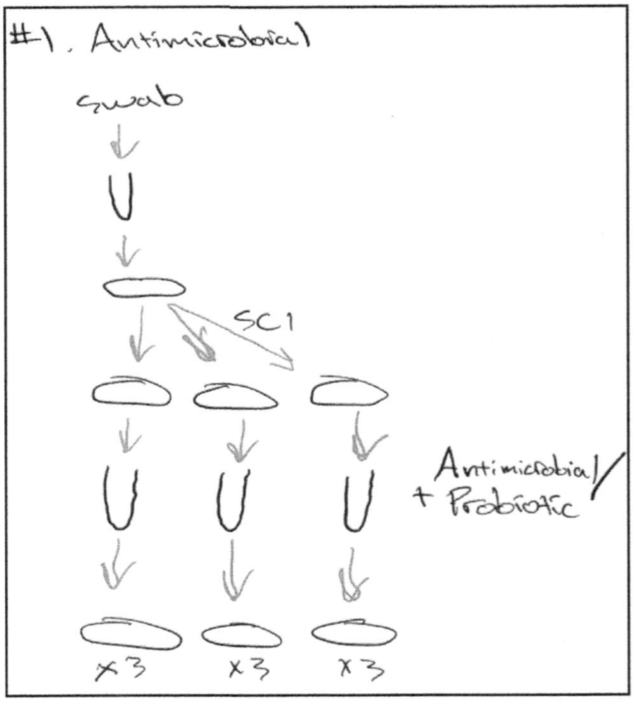

Figure 2. A drawn-out overview of our planned methods for this experiment

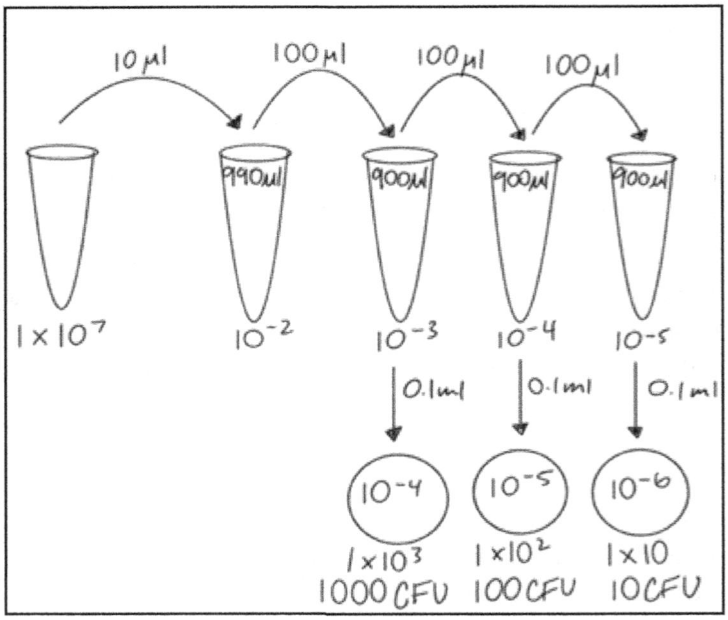

Figure 3. A drawn-out overview of our serial dilution

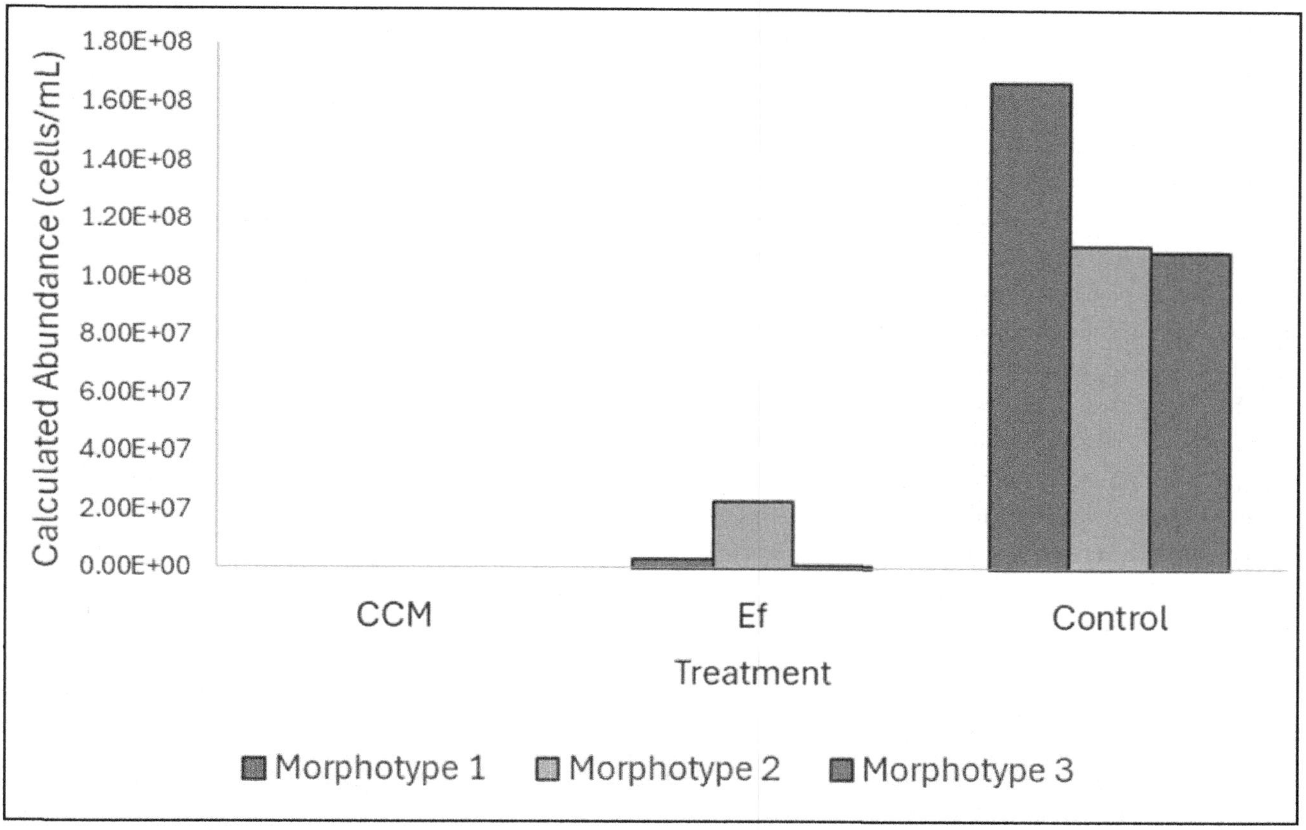

Figure 4. Individual values of calculated concentrations of the CCM treatment, the inoculation with *E. faecalis*, and the controls. Each bar represents CFUs on individual plates. Results indicate that CCM treatment completely inhibited each morphotype's growth and was impacted to different degrees by *E. faecalis* treatment.

absorbance and the Agilent E. coli cell culture concentration from the OD600 calculator, we obtained a rough estimate of the concentration of our control broths to be about 1x10^7 from a reading of 0.01 of the absorbance at 600nm. We then used this estimate to form our dilution scheme (Fig. 3). Then, 0.1 ml of each dilution was used to create an LB spread plate, which was estimated to contain 1000, 100, and 10 CFUs (Fig. 3). This process was repeated with the treated broths. The spread plates were incubated at 37°C for 24 hours. Following incubation, we performed a CFU count of each spread plate to calculate the initial concentration of each broth.

Initial data was organized in a Google spreadsheet and then imported into an Excel spreadsheet. We performed a one-way ANOVA on each 10^-4 dilution treatment plate to identify significant differences in data between CCM treatments, E. faecalis treatments and our controls. Data was presented as a bar graph. (Fig. 4)

RESULTS

We cultured a total of three unique bacteria from the canine oral sample. Each bacterial morphotype was treated as a replicate for each treatment. The spread plates from the CCM treatment produced zero colony-forming units at every dilution for every bacterium, which concluded in a calculation of zero bacterial abundance in the treatment solutions. (Fig. 5) The treatment tubes inoculated with both the oral bacteria and *E. faecalis* were calculated to have a concentration after culturing of the oral bacteria at 3,000,000 cells/mL for morphotype 1, 23,000,000 cells/mL for morphotype 2, and 1,000,000 cells/mL for morphotype 3. The control concentrations were 167,000,000 cells/mL for morphotype 1, 111,000,000 cells/mL for morphotype 2, and 109,000,000 cells/mL for morphotype 3. We found a significant difference in the abundance of bacteria between the different treatments and the control ($df=2$, $F=37.89$, $P<0.05$)

DISCUSSION

Our experiment highlighted the effectiveness of CCM in limiting the growth of bacteria. We hypothesized that *E. faecalis* would inhibit the growth of some bacteria while still allowing for the growth of other bacteria. While we hypothesized that CCM treatment would inhibit the growth of the dog microbiome, independent of the type of bacteria from the dog's mouth. Upon implementation of our experiment, we found that both treatments decreased the growth of isolated bacterial strains when compared to our controls.

A limitation of our experiment is that we did not identify the bacterial strains that we isolated. We instead determined the bacterial strains were different based on phenotype, which was later corroborated based on the different growth of the bacteria in the treatment with the E. faecalis. For the probiotic treatment, the addition of probiotic likely increased the competition for nutrients for each of the strains, thus resulting in less growth than in the isolated control treatments.

A particular area of interest in our results shows that the rates of growth relative to the other bacterial morphotypes changed between the probiotic treatment and the control treatment (Fig. 4). This is highlighted by the fact that in the control, bacterial morphotype 1 shows the most amount of growth out of the three. With the addition of the probiotic, however, this changed, and bacteria morphotype 2 exhibited substantially higher proliferation compared to morphotype 1. This could suggest that there is an interactive effect between treatment and bacterial species. We did not run replicates on each of the morphotypes, so this finding is purely off of the data we collected from running it once. This could indicate that the probiotic could be a valuable tool in adjusting the microbiome of the mouth by selectively promoting growth conditions for one bacterial species over another. In contrast, treatment with CCM resulted in complete inhibition of bacterial growth under the tested conditions.

Had the treatment been applied to pre-established colonies, it is possible that some bacterial strains would have survived, which would have given us better insight into how CCM interacts with the oral microbiome of a canine. However, since we implemented the treatment on the inoculation day, that may be the reason we did not see any growth. We used industry standard values for implementation of the antimicrobial concentrations, but it could also be possible that the concentration needs to be diluted to see if the CCM affects growth differently depending on what bacterial strain it is used to treat. Additionally, it should be looked into how CCM affects the microbiome of the gut bacteria. If it does inhibit the growth of bacteria and is not selective for harmful or beneficial bacteria, our results indicate that it is possible that CCM could potentially destroy the bacteria of the mouth and gut. This would make the use of CCM in water additives potentially negative, as CCM is ingested as part of treatment and would make its way into the mouth and gut of the dogs. Future research for looking into the use of this probiotic and antimicrobial should include treatments of pre-established colonies, as well as treating the bacteria in situations where the bacterial species are grown together.

In conclusion, both the CCM and *Enterococcus faecalis*

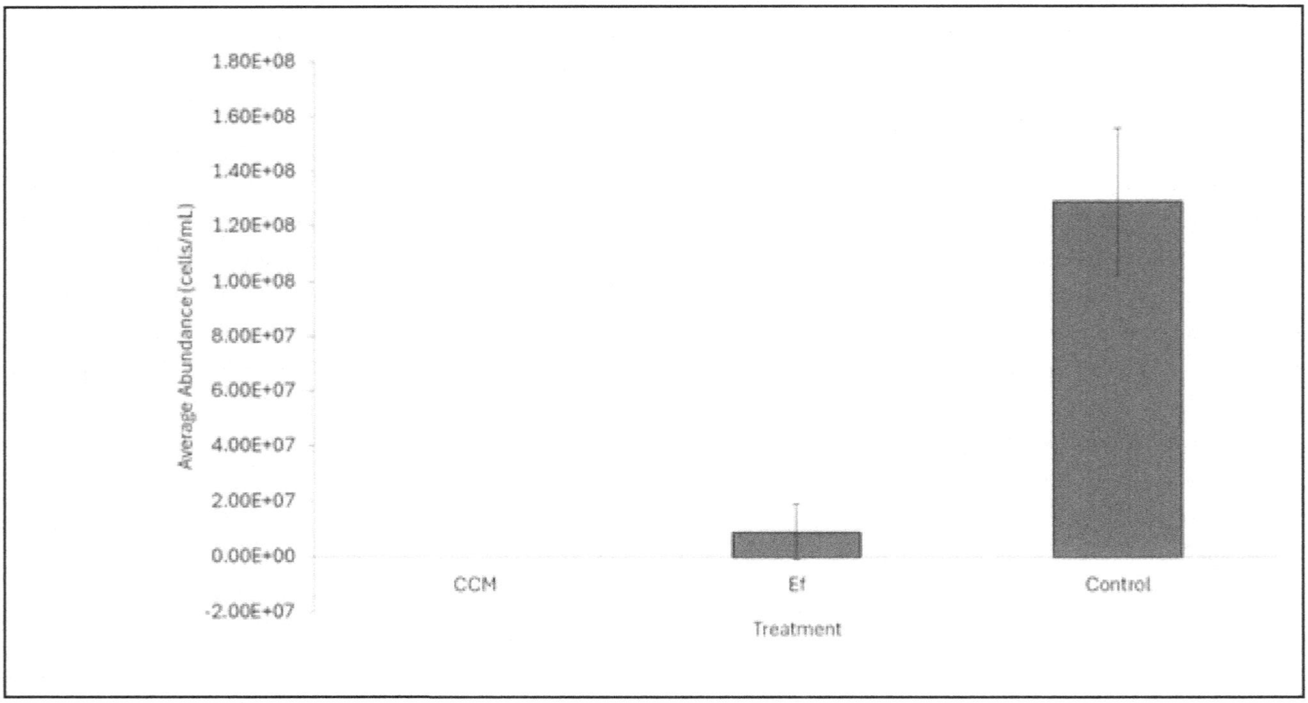

Figure 5. Average values and standard deviation of calculated concentrations of the CCM treatment, the inoculation with *E. faecalis*, and the controls. Results indicate that CCM completely inhibits the growth of bacteria while *E. faecalis* reduces growth among all morphotypes without completely inhibiting growth.

show promising results on being able to adjust the microbiome of the mouth. Also, running replicates of the experiment would give a more conclusive insight into whether or not we did see an interactive effect and would allow us to run a two-way ANOVA. Implementation of both CCM and Enterococcus faecalis must be taken into consideration, as both have the potential to affect other microbiomes.

LITERATURE CITED

1. Segata, N., Haake, S.K., Mannon, P. et al. 2012. *Composition of the adult digestive tract bacterial microbiome based on seven mouth surfaces, tonsils, throat and stool samples.* Genome Biol 13, R42
2. Whitmore SE, Lamont RJ 2014. *Oral Bacteria and Cancer.* PLoS Pathog 10(3): e1003933.
3. Kaur S, White S, Bartold PM. 2013. Periodontal Disease and Rheumatoid Arthritis: A Systematic Review. *Journal of Dental Research.* J Dent Res 92(5): 399-408
4. Elliott DR, Wilson M, Buckley CMF, Spratt DA 2005. Cultivable Oral Microbiota of Domestic Dogs. J Clin Microbiol43:.
5. Niemiec, B. A., Gawor, J., Tang, S., Prem, A., and Krumbeck, J. A. (2021). The bacteriome of the oral cavity in healthy dogs and dogs with periodontal disease. *Am. J. Vet. Res.* 83, 50–58. doi: 10.2460/ajvr.21.02.0027
6. Pilla R, Suchodolski JS. 2021 The Gut Microbiome of Dogs and Cats, and the Influence of Diet. Vet Clin North Am Small Anim Pract. (3):605-621. doi: 10.1016/j.cvsm.2021.01.002. Epub 2021 Feb 27. PMID: 33653538.
7. Bell SE, Nash AK, Zanghi BM, Otto CM, Perry EB. 2020 An Assessment of the Stability of the Canine Oral Microbiota After Probiotic Administration in Healthy Dogs Over Time. Front Vet Sci.7:616. doi: 10.3389/fvets.2020.00616. PMID: 33062653; PMCID: PMC7517700.
8. Kislik G, Zhou L, Rubbi L, Pellegrini M. 2024 Age-correlated changes in the canine oral microbiome. Front Microbiol. 15:1426691. doi: 10.3389/fmicb.2024.1426691. PMID: 39081893; PMCID: PMC11287893.
9. Teng F, He T, Huang S, Bo C-P, Zhen L, Chang J-L, Liu J-Q, Charbonneau D, Xu J, Li R, Ling J-Q. 2016 Cetylpyridinium chloride mouth rinses alleviate experimental gingivitis by inhibiting dental plaque maturation. *Int J Oral Sci.* 8:182-190.
10. Oba PM, Carroll MQ, Alexander C, Somrak AJ, Keating SCJ, Sage AM, Swanson KS. 2021 Dental chews positively shift the oral microbiota of adult dogs. J Anim Sci.99(7):skab100. doi: 10.1093/jas/skab100. PMID: 33780530; PMCID: PMC8257028.
11. Sharma N, Bhatia S, Sodhi AS, Batra N. 2018 Oral microbiome and health. AIMS Microbiol. 4(1):42-66. doi: 10.3934/microbiol.2018.1.42. PMID: 31294203; PMCID: PMC6605021.
12. Yamada T, Takahashi-Abbe S, Abbe K. 1985. Effects of oxygen on pyruvate formate-lyase in situ and sugar metabolism of Streptococcus mutants and Streptococcus sanguis. Infect Immun 47:.
13. Thongma N, Sivamaruthi BS, Bharathi M, Tansrisook C, Peerajan S, Tanongpitchayes K, Chawnan N, Rashmi S, Thongkorn K, Chaiyasut C. 2023 Influence of Gallic Acid-Containing Mouth Spray on Dental Health and Oral Microbiota of Healthy Dogs: A Pilot Study. *Veterinary Sciences.* 10(7):424.
14. M.S. Castro, M.A. Molina, M.B. Azpiroz, A.M. Díaz, R. Ponzio, M.D. Sparo, M.A. Manghi, A.M. Canellada, 2016 Probiotic activity of *Enterococcus faecalis* CECT7121: effects on mucosal immunity and intestinal epithelial cells, *Journal of Applied Microbiology*, Volume 121, Issue 4, Pages 1117–1129
15. Amira A Bhalodi, Tjitske S R van Engelen, Harjeet S Virk, W Joost Wiersinga, 2019 Impact of antimicrobial therapy on the gut microbiome, *Journal of Antimicrobial Chemotherapy*, Volume 74, Issue Supplement_1,, Pages i6–i15,
16. Benyacoub J, Cavadini C, Sauthier T, Schiffrin EJ, von der Weid T, Czarnecki-Maulden GL, Anderson RE. 2003 Supplementation of Food with Enterococcus faecium (SF68) Stimulates Immune Functions in Young Dogs. *The Journal of Nutrition.* 133(4): 1158-1162.
17. Jeeho Im E, Hyun-Yup Lee H, Kim M, Kim M-K. 2023. Evaluation of Enterococcal Probiotic Usage and Review of Potential Health Benefits, Safety, and Risk of Antibiotic-Resistant Strain Emergence. *Antibiotics.* 12(8):1327.
18. Shi D, Mi G, Wang M, Webster TJ. 2018 In vitro and ex vivo systems at the forefront of infection modeling and drug discovery. *Biomaterials.* 198:228–249.
19. Doke SK, Dhawale SC. 2015 Alternatives to animal testing: A review. *Saudi Pharm J.* 23(3):223-9.

Humboldt Journal of
Microbiology
2025, Volume 24

Growth Trends in Abundance and Diversity of Synthetic Microbial Communities over Varying Thermal Conditions

Mariska Kessler, Jocelyn Wolfinger, Destiny Alcaraz

ABSTRACT

Microorganisms create various communities that act as the foundation of ecosystems. Abiotic factors such as temperature variability and environmental fluctuations are known to produce shifts in the abundance and diversity of microbial communities, but our current understanding is limited. In this study, we aimed to address this gap by observing abundance and diversity growth patterns under different temperature conditions within individual species and communities. Samples of 4 isolates and 4 artificial communities were grown at constant and fluctuating temperatures. Results found that the isolate samples had the highest abundance under constant high temperatures, whereas the community samples had the highest abundance under fluctuating temperatures. The highest diversity was found in the community samples grown in a constant low temperature. These findings suggest that microbial community makeup can be altered by growth temperature. Community samples exhibit the highest growth and diversity under conditions that more closely resemble their natural temperature environment. Researchers should account for this factor when designing experiments with diverse microbial populations.

INTRODUCTION

Microorganisms form distinct and diverse communities that are essential for ecological processes such as nutrient cycling by breaking down organic materials, enhancing disease resistance by outcompeting pathogens, and contributing to ecosystem stability, especially in dynamic environments (1). The structure of a microbial community is often closely tied to its function within the environment or host, like the symbiotic relationship between Bacteroidetes and Firmicutes found within the human gut microbiome (2). Microbial communities typically display unimodal diversity and abundance patterns over time, where diversity and abundance increase up to a certain point before declining due to resource limitations or species interaction (3). This diversity is shaped by a variety of factors, interactions with the host organism, and long-term evolutionary pressures4. Research shows that even small changes in environmental conditions can have significant impacts on microbial community structure and function (1).

Temperature fluctuations are especially important when considering how microbes interact and coexist within communities. Each organism has a thermal performance range where it functions most optimally (5). Deviations from this range can affect cell growth and metabolic processes in microorganisms, impacting species interactions and ultimately shaping a community's size and dynamics (6). Fluctuations in temperature can cause periodic shifts in abundance and diversity, allowing some species to persist whereas others decline (7). Research suggests that these fluctuations play a key role in the structure of microbial communities over time, making them a critical factor in ecosystem development (8). Specifically, there is limited understanding of how species interactions, diversity, and community dynamics are maintained under fluctuating thermal regimes outside of controlled, host-associated systems. This knowledge gap represents a critical challenge in microbial ecology, as natural temperature variability is a pervasive and influential factor in many ecosystems.

Darlingtonia californica, also known as the Cobra Lily, is a carnivorous plant from North America with a modified pitcher-like leaf that hosts a diverse community of microorganisms, including bacteria, fungi, and other microbes that thrive in nutrient-poor condition (7). The pitcher of these plants traps insects, creating a unique enclosed microhabitat for microbes to break down organic matter and recycle nutrients that are essential for the plant's overall function (8). One study demonstrates that bacteria within *D. californica* pitchers assist with prey retention by altering the physical properties of the pitcher's fluid; enhances the plant's ability to capture and digest prey (7). Furthermore, studies on related pitcher plants, such as *Sarracenia purpurea*, have shown that bacterial communities within the pitchers are diverse and changing over time, indicating a dynamic interaction between the plant and its microbiome (9). These characteristics make *D. californica* an excellent model for studying microbial ecology and assessing how environmental factors, such as natural thermal fluctuations, impact the abundance and diversity of bacterial communities. However, despite the insights from studies on pitcher plants, there remains a broad gap in our understanding of how bacterial communities coexist and function in fluctuating environments beyond these specialized systems.

Temperature fluctuations are likely a major determinant of bacterial abundance and diversity in *D. californica*. Previous studies have explored how natural temperature fluctuations, specifically climate change-related fluctuations, affect species survival rates, shift species distributions, or create mismatched community interactions, which overall affect the functionality of the community and the organism (6). Other studies delve into how temperature fluctuations affect bacterial interactions, such as the competition and coexistence of species, and found that some species need fluctuating conditions to function and cannot coexist at certain constant temperature ranges (7). While both areas of research show temperature-dependent competitive dynamics, the broader understanding of how bacterial communities respond to fluctuating versus constant temperatures, especially under conditions that involve *D. Californica's* natural habitat, remains limited.

We aim to determine whether temperature fluctuations over time shift the makeup of bacterial communities. Specifically, if growth at fluctuating temperatures have a greater impact on bacterial community abundance and diversity than at constant temperatures. We expect this pattern to hold in our system, as it reflects the natural thermal variability that *D. californica* experiences. In laboratory settings, these types of organisms are often grown at constant temperatures, which is not reflective of their natural conditions and may lead to misleading conclusions regarding bacterial abundance and diversity. If temperature fluctuations promote microbial coexistence, we would expect to see higher microbial diversity and abundance in the growth of fluctuating environments compared to constant temperatures. If fluctuating temperatures were to introduce a disturbance in growth, then community stability would be reduced, leading to lower diversity and abundance.

MATERIALS AND METHODS

The effect of temperature fluctuations on the diversity of bacterial communities in *Darlingtonia californica* will be measured using 4 morphologically distinct species collected from the pitcher fluid: *Bacillus mycoides, Arthrobacter gandavensis 1, Microbacterium oxydans, and Leucobacter sp. dR13-9.* These isolates were obtained from the pitchers of *D. californica* plants collected in three locations: Gasquet, CA, Shasta, CA, and Quincy, CA. The isolates were removed from cryopreservation, revived (DMSO Cryopreservation Protocol, CCG), and cultured in LB broth for 24 hours. After incubation, isolates were plated at 30°C for 24 hours to ensure the correct isolates were present, confirm culture purity, and create a list of identifiable colony morphologies to later identify taxa in community samples (Figure 1).

We created four different communities using combinations of three isolates each, with three replicates per group. Community A will consist of Isolate 1 (*B. mycoides*), Isolate 2 (*A. gandavensis*), and Isolate 3 (*M. oxydans*). Community B will consist of Isolate 1, Isolate 2, and Isolate 4 (*Leucobacter sp.*). Community C will consist of Isolate 2, Isolate 3, and Isolate 4. Community D will consist of Isolate 1, Isolate 3, and Isolate 4. Each isolate was also grown separately under the same conditions and in triplicate. All cultures were grown in LB broth and subjected to three temperature treatments using a New Brunswick Scientific I 24 Incubator Shaker at 150 RPMs. Based on environmental data from collection sites and incubator constraints, high and low temperatures of 35 °C and 25 °C were designated as the bounds of the experiment 10.

Samples were incubated under three conditions. The conditions included a constant high temperature (35 °C for 24 hours), a constant low temperature (25 °C for 24 hours), and a fluctuating temperature cycle (6 hours at 25 °C, 12 hours at 35 °C, and 6 hours at 25 °C). After incubation, we measured bacterial abundance using both direct cell counts and spectrophotometry from the first replicate of each isolate and community broth. Based on these measurements, we designed a dilution scheme to achieve about one hundred colony-forming units per plate for diversity quantification of community samples. All replicates were serially diluted, plated on LB agar using the spread plate method, and incubated at 30 °C for 48 hours. Colony types were identified by comparing with reference plates and used to quantify the diversity of communities.

To analyze our data, we performed a variety of tests. A two-way ANOVA was done on our abundance data between communities and isolates to assess the effects of temperature treatment (high, low, fluctuating) and bacterial type (isolates vs communities), and between each individually. We also did separate one-way ANOVAs to determine the effect of temperature on abundance within isolates and communities independently. For diversity, we utilized Shannon Diversity Index, richness, and evenness statistical tests, as well as one-way

ANOVAS to test for differences in diversity across temperature treatments for each taxon. Finally, Tukey's post-hoc tests were done after each test to identify pairwise differences among treatments.

Due to time constraints, initial spectrophotometric readings were only obtained for one replicate of each sample immediately following incubation. To preserve the remaining samples, all were refrigerated at 4°C to minimize further bacterial activity. Spectrophotometric readings for all samples were taken 48 hours post-refrigeration. Using the difference between the pre- and post-refrigeration reading from the first replicates, we calculated the correction ratio and applied it to the remaining replicates of the same community or isolate to normalize the data across time points. This accounted for the proportional decrease in absorbances over time due to refrigeration. We observed a consistent decrease in absorbances across all samples in the refrigeration, making this adjustment necessary for normalizing all data across all treatments.

During the experiment, *Pseudomonas aeruginosa*, a gram-negative bacillus commonly found in *D. californica* pitcher fluid, was detected as a contamination in both broth cultures and dilution plates. Although the exact source of contamination is unknown, *P. aeruginosa* was consistently present across all samples. Given its ecological relevance and even distribution, we chose to include it in our community analyses as a natural component of the *D. californica* microbiome.

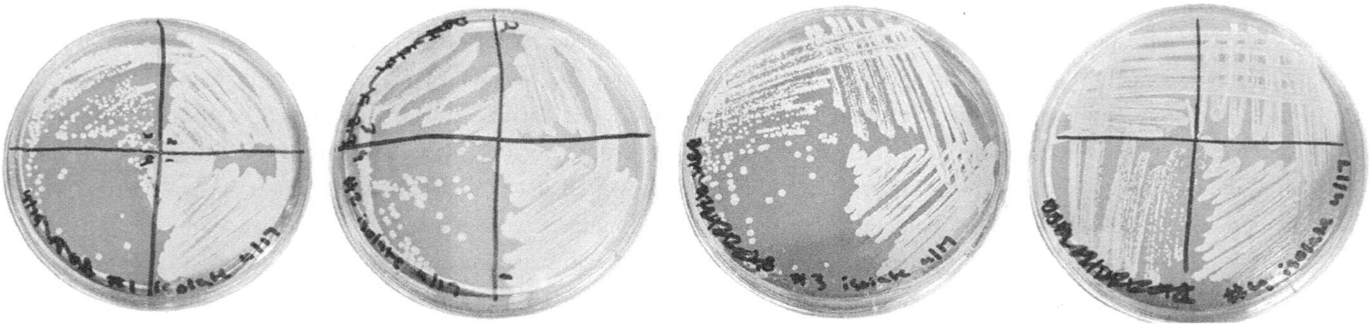

Figure 1. Colony morphology of the four isolates we initially started with, from left to right: Bacillus mycoides, *Arthrobacter gandavensis* 1, *Microbacterium oxydans*, and Leucobacter sp. dR13-9. These will serve as references when initially reviving the isolates and for distinguishing isolates from each other when we plate communities to determine diversity.

RESULTS

Our first statistical analysis aimed to determine the effect of temperature and bacterial composition on abundance. A two-way ANOVA assessed how temperature treatment (low, high, fluctuating) and bacterial composition (isolates vs. communities) influenced bacterial abundance. Both temperature ($F(2, 48) = 237.78$, $p < 0.001$) and bacterial type ($F(7, 48) = 392.89$, $p < 0.001$) had significant effects on abundance, as well as their interaction ($F(14, 48) = 37.57$, $p < 0.001$). Low temperature treatments consistently had the lowest bacterial abundance across both isolates and communities (Figure 2). Conversely, communities had higher abundances under fluctuating temperatures and isolates responded best at high temperatures (Figure 2 and Figure 3). Tukey's test following ANOVA results revealed significant differences between low and high treatments along with low and fluctuating treatments, in both isolates and communities.

A one-way ANOVA on isolate abundance found that Isolate 2 (*Arthrobacter gandavensis*) was persistently the most abundant at high and fluctuating temperatures ($F(2, 33) = 3.56$, $p = 0.0397$). In contrast Isolate 3 (*Microbacterium oxydans*) had the lowest abundance across all conditions, with the only significant difference in growth between the low and high temperature treatments (Tukey's post hoc: $p = 0.037$) (Figure 2 and Figure S1.).

Community-specific differences revealed that Community B had significantly higher growth than other communities at all temperatures ($F(2, 33) = 3.19$, $p = 0.0542$) (Figure 3). Other communities showed more variable, but generally lower abundance (Figure S2). A one-way ANOVA was conducted to provide additional insight into Community B's performance, revealing the reduced presence of other isolates when Isolate 3 was present (Figure S3). For comparison, the same analysis was run for the rest of the communities to identify prevalent or absent species (Figures S4-S6).

The Shannon Diversity Index was used to evaluate how temperature influenced community richness and evenness. Diversity was significantly higher at low temperatures than at high temperatures ($F(2, 33) = 8.36$, $p = 0.0012$), and fluctuating temperatures produced intermediate levels of diversity (Figure 4). Richness followed this same trend, with low temperature samples having the most unique taxa ($F(2, 33) = 10.56$, $p = 0.00029$) (Figure S7). Evenness was also greater at low temperatures, indicating species were more evenly distributed ($F(2, 12) = 85.86$, $p < 0.0001$) (Figure S8).

In visual plate observations, the production of pyocyanin, a green-blue pigment produced as a virulence factor by *Pseudomonas aeruginosa*, was increased in community samples from low and fluctuating temperatures, compared to high temperatures; qualitatively supporting diversity-abundance trends (Figure 5).

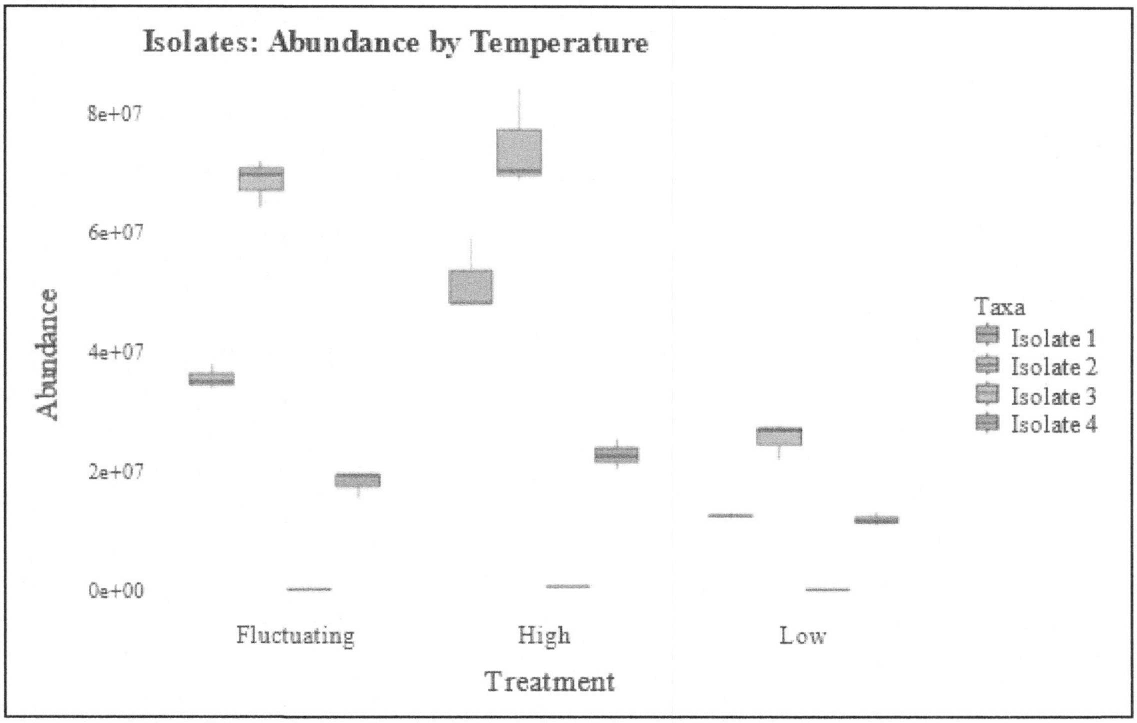

Figure 2. Boxplot of bacterial abundance for individual isolates across temperature treatments. Each box represents the distribution for one isolate under a certain temperature. This shows isolate 2 being the most dominant in terms of abundance at fluctuating and high temperatures.

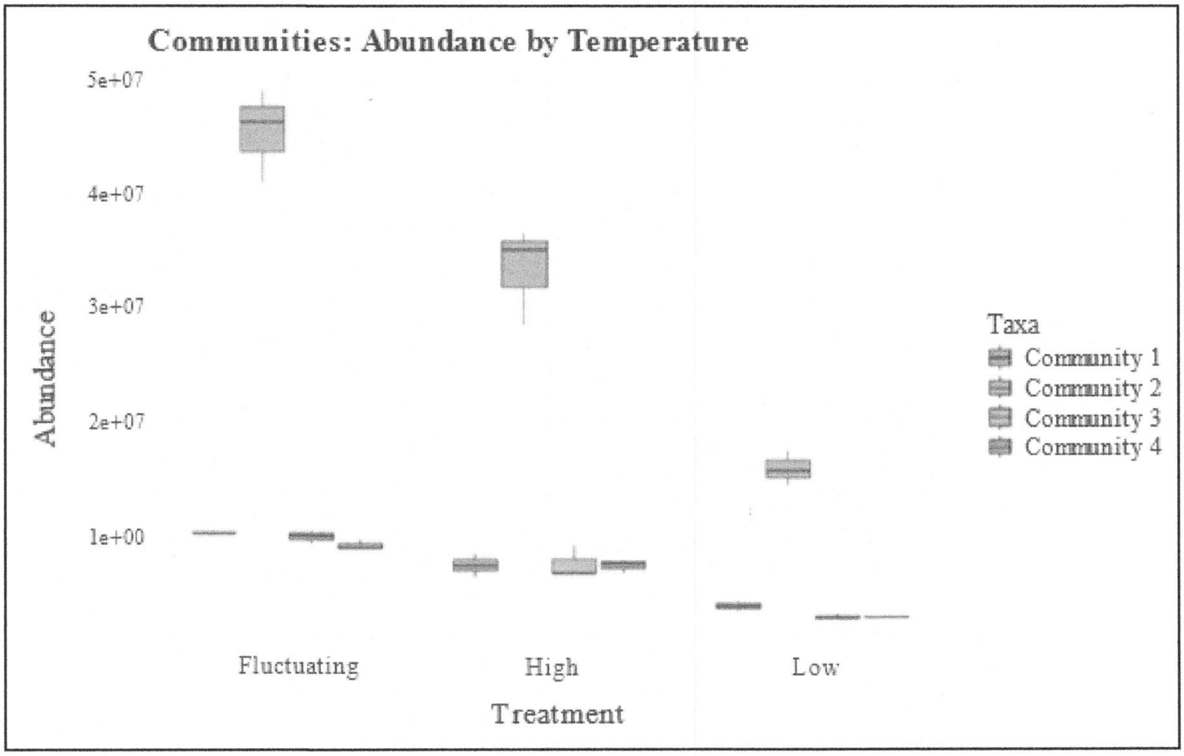

Figure 3. Boxplot of microbial abundance for different community types across temperature treatments. Each box represents abundance for a specific community at a given temperature.

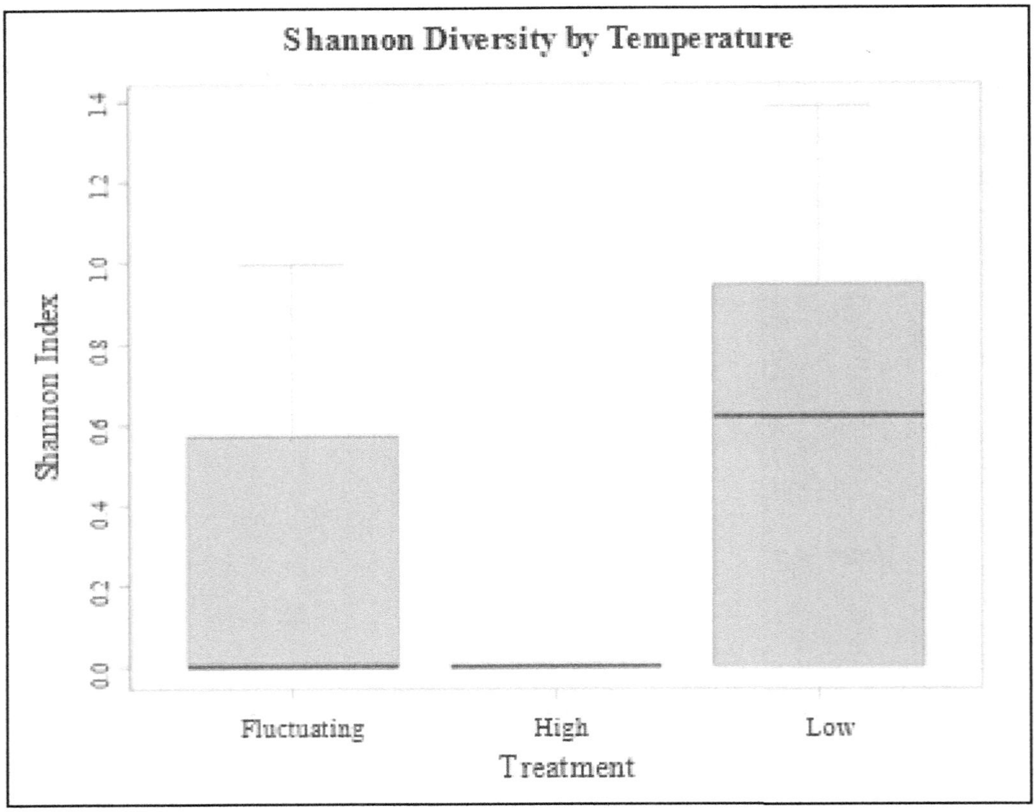

Figure 4. Boxplot showing Shannon diversity of bacterial communities under three temperature regimes: fluctuating, high, and low. Shannon index accounts for both richness and evenness. Communities grown at low temperatures exhibited the highest diversity, while those at high temperatures resulted in moderate diversity.

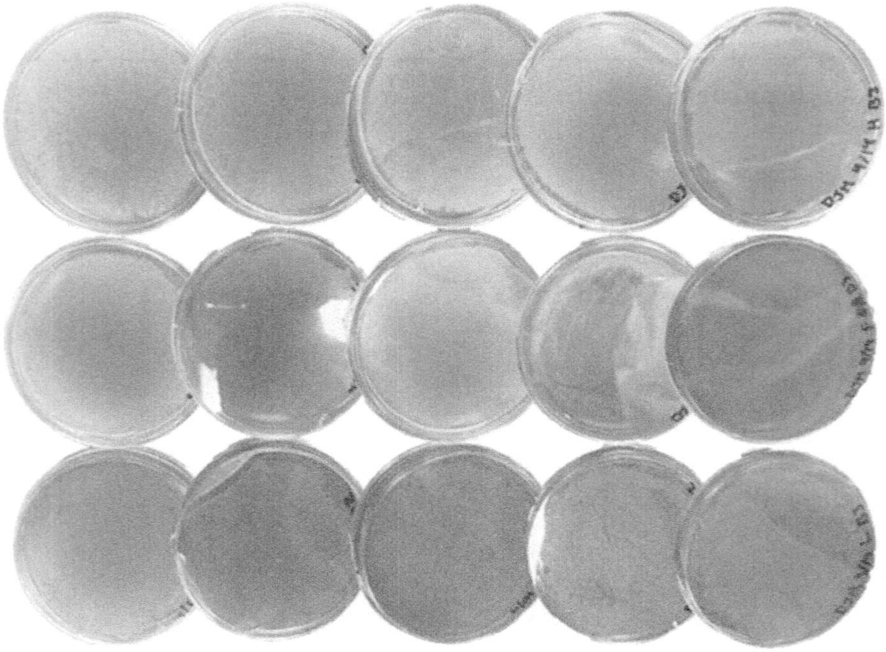

Figure 5. Plates from serial dilutions done on Communities A-D represent the distribution of pyocyanin pigment production across different temperature treatments. Communities exposed to the low temperature treatment are on the bottom row, communities exposed to the fluctuating treatment are in the middle row, and communities exposed to the high temperature treatment are on the top row. Each community had slight variations in the amount of pigment produced, but we see a trend of more pyocyanin in the low and then fluctuating temperatures, with the least amount of pigment in the high temperature treatment communities.

DISCUSSION

The results of our experiment support our hypothesis that temperature and community composition significantly affect abundance and diversity; fluctuating temperatures promote greater abundance in communities and lower temperatures promote greater diversity. We found that these two factors interact, as different taxa had varying responses to the same temperature treatments. This interaction displays the complexity of microbial ecology, with both environmental conditions and community composition shaping outcomes. We observed significant change in bacterial abundance with temperature, but the patterns of abundance were differed between isolate and community samples. Isolate samples had significantly higher abundances when exposed to constant high temperatures compared to constant low temperatures, but no significant difference between fluctuating and low temperature conditions. In contrast, communities had significantly higher abundances in fluctuating conditions than in low temperatures, with no difference between constant high and low temperatures. This difference shows that individual taxa rarely deviate from their optimal thermal ranges, as opposed to community interactions that can respond to more dynamic temperature variability. This is likely because fluctuating temperatures mediate species interactions, allowing temporal shifts in competitive dynamics or coexistence that would not otherwise occur under constant conditions.

We obtained an unexpected result from this study regarding artificial Community B; significantly higher growth compared to other communities. This community included Isolate 1 (*Bacillus mycoides*), Isolate 2 (*Arthrobacter gandavensis* 1), Isolate 4 (*Leucobacter sp. dR13-9*), and Pseudomonas aeruginosa. All other communities, which included Isolate 3 (*Microbacterium oxydans*), had similarly lower growth. This could indicate that Isolate 3 is a strong competitor, preventing other taxa from growing in a community setting. Moreover, when we grew Isolate 3 independently, it had the lowest growth of the Isolates. This suggests that its competitive advantage in communities is a result of interference competition, where it's presence may directly impede the growth of other species. This interaction highlights the importance of species interactions and biotic effects, where certain taxa, despite low individual growth, can suppress the growth or function of other community members. Such effects may occur through mechanisms such as resource competition, chemical inhibition, or interference with cooperative processes (12).

The Shannon Diversity, richness, and evenness analyses indicated that constant low temperatures supported the most diverse communities. In contrast, constant high temperatures were associated with lower diversity, likely due to dominance by a few heat-tolerant species (for instance, *P. aeruginosa*). Finally, fluctuating temperatures produced an intermediate diversity, possibly due to periodic shifts that allowed fewer heat-tolerant competitors to persist. These results suggest that cooler or fluctuating temperatures are better capable of hosting diverse bacterial communities, whereas constant high temperatures reduce diversity, possibly because only heat-tolerant taxa like *P. aeruginosa*, can persist under such conditions.

Although *Pseudomonas aeruginosa* was not intentionally included in our community samples, it offered meaningful results regarding diversity. This species can be commonly found within the pitchers of *Darlingtonia californica* and was therefore incorporated into our statistical analysis of the communities (12). The contamination was likely from P. aeruginosa cultured in the same lab. All agar plates with P. aeruginosa had pyocyanin, a blue-green virulence factor pigment, and showed observable trends that could potentially serve as a physical marker of metabolic activity (13). On serial dilution spread plates we visually observed that pyocyanin was most prominent in low and fluctuating temperature treatments, whereas high temperature treatments produced little to no visible pigment. This trend indicates that higher bacterial diversity may promote secondary metabolite production such as pyocyanin (14). Pyocyanin production may be linked to interspecies competition or signaling, as its presence often increases when *P. aeruginosa* interacts with other species (15). This visual, qualitative evidence supports the hypothesis that community structure influences ecological function.

Our findings suggest that natural thermal fluctuations, like those experienced in *D. californica* pitchers, may promote greater bacterial coexistence and higher diversity and functional output. We observed more pyocyanin production on community agar plates that underwent the constant low and fluctuating temperatures treatments, with fluctuating treatments also exhibiting the highest abundance and diversity. This highlights the importance of temperature variability in maintaining bacterial community composition. This study demonstrates that increased variability may be beneficial when considering the broader context of climate change, where extreme temperatures are becoming more frequent. In contrast, prolonged exposure to high temperatures could reduce diversity and therefore ecosystem function.

While our study provides some meaningful explanations for community interactions in *D. californica*, we encountered limitations. A notable constraint of this study was the use of a simplified community system with four isolates, compared to natural bacterial communities in *D. californica* pitchers, which contain hundreds of bacterial, fungal, and protist species. This simplification fails to account for the complexity found within these communities, as well as the natural interactions between microorganisms in this microbiome, such as competition and predation. An additional constraint was the limited number of experimental trials due to time constraints. Ideally, repeated trials of this experiment would be performed to confirm trends and increase the robustness of our statistical analyses. Lack of temporal replication limits our ability to assess how communities might shift or stabilize over time under varying temperature conditions.

While this study provides several insights regarding temperature fluctuations impacting bacterial diversity and abundance, considerable gaps in knowledge remain, and further experimentation is needed. The investigation of long-term effects of fluctuating temperatures on community structure could provide a better understanding of this community's ecological function. It may be beneficial to use molecular methods to determine the mechanisms underlying bacterial inhibition as well as evidence of how temperature influences gene expression and metabolic production. It would be beneficial to explore a wide range of functional traits beyond pigment production, like enzyme activity, nutrient cycling roles, or resistance traits.

Overall, our findings support the hypothesis that temperature variability can support greater bacterial diversity and function compared to constant thermal conditions. This is especially important when considering systems such as *D. californica* and its relevance in ecology research. By mimicking the natural thermal variability of its pitcher microbiome, we observed patterns in abundance, diversity, and metabolite production that reflect ecological responses in situ. Laboratory studies often incubate such organisms under constant temperatures, failing to replicate the thermal variability experienced in nature. As a result, experiments may produce misleading conclusions about bacterial abundance and diversity in systems adapted to fluctuating conditions. In the context of climate change, understanding these dynamics becomes even more critical, as systems like *D. californica* may rely on thermal variation to maintain stable microbiomes. Loss of variation due to constant high temperatures associated with global warming, may alter community structure and function in detrimental ways. This work highlights how model systems like *D. californica* can aid in studying complex ecological processes. While limited in scale and taxonomic diversity, our findings provide a foundation for future research into the responses of bacterial community's to varying temperature conditions, as well as uncovering the mechanisms behind temperature-mediated communities and their broader ecological implications.

LITERATURE CITED

1. Robinson CJ, Bohannan BJM, Young VB. 2010. From Structure to Function: The Ecology of Host-Associated Microbial Communities. Microbiol Mol Biol Rev74. https://doi.org/10.1128/mmbr.00014-10.
2. Giri S, Shi H, Typas A, Huang KC. 2025. Harnessing gut microbial communities to unravel microbiome functions. Curr Opin Microbiol. 83:102578. https://doi.org/10.1016/j.mib.2024.102578.
3. Allison SD, Martiny JBH. 2008. Resistance, resilience, and redundancy in microbial communities. Proc Natl Acad Sci U S A. 105(Suppl 1):11512–11519. https://doi.org/10.1073/pnas.0801925105.
4. Larsen PE, Dai Y. 2020. The metabolome of the human gut microbiome is predictive of host dysbiosis. mSystems. 5(5):e00738-20. https://doi.org/10.1128/mSystems.00738-20.
5. Rebolledo AP, Argo CM, Monro K. 2021. Thermal Performance Curves Are Shaped by Prior Thermal Environment in Early Life. Frontiers in physiology. 12, 738338. https://pmc.ncbi.nlm.nih.gov/articles/PMC8564010/.
6. Zander A, Bersier LF, Gray SM. 2017. Effect on temperature variability on community structure in a natural microbial food web. Global Change Biology, 23(1), 56-67. https://onlinelibrary.wiley.com/doi/10.1111/gcb.13374
7. Jiang L, Morin PJ. 2007. Temperature fluctuation facilitates coexistence of competing species in experimental microbial communities. J Anim Ecol. 76(4):660–668. https://doi.org/10.1111/j.1365-2656.2007.01252.x.
8. Lara-Gutierrez J, Nguyen J, Stocker R. 2021. Environmental fluctuations and their effects on microbial communities, populations, and individuals. FEMS Microbiology Reviews, 45(4). https://academic.oup.com/femsre/article/45/4/fuaa068/6041721.
9. Melchior PP, Reiss E, Payne Z, Vuong N, Hovorka K, Lindsay HL, Diaz GR, Gaire T, Noyes N. 2024. Analysis of the northern pitcher plant (Sarracenia purpurea L.) phytotelm bacteriome throughout

a temperate region growing season. PLOS One, Volume 19, Issue 7. https://journals.plos.org/plosone/article?id=10.1371/journal.pone.0306602.
10. NOAA (National Oceanic and Atmospheric Administration). NOAA homepage. https://www.noaa.gov/.
11. Kanehisa M, Goto S, Furumichi M, Tanabe M, Hirakawa M. 2010. KEGG for representation and analysis of molecular networks involving diseases and drugs. Nucleic Acids Res. 38(Database issue):D355–D360. https://doi.org/10.1093/nar/gkp896.
12. Armitage DW. 2016. Microbes in time: Incorporating bacteria into ecosystem development theory. Order No. 10150912, University of California, Berkeley, 2016. https://www.proquest.com/docview/1834519692?pq-origsite=gscholar&fromopenview=true&sourcetype=Dissertations%20&%20Theses.
13. Leibniz Institute DSMZ–German Collection of Microorganisms and Cell Cultures. BacDive: Bacterial diversity metadatabase. Pseudomonas aeruginosa strain entry 12800. https://bacdive.dsmz.de/strain/12800.
14. Hall S, McDermott C, Anoopkumar-Dukie S, McFarland AJ, Forbes A, Perkins AV, Davey AK, Chess-Williams R, Kiefel MJ, Arora D, Grant GD. 2016. Cellular effects of pyocyanin, a secreted virulence factor of Pseudomonas aeruginosa. Toxins. 8(8):236. https://doi.org/10.3390/toxins8080236.
15. Zhao J, Xie X, Jiang Y, Li J, Fu Q, Qiu Y, Fu X, Yao Z, Dai Z, Qiu Y, Chen H. 2023. Effects of simulated warming on soil microbial community diversity and composition across diverse ecosystems. Sci Total Environ. 900:165865. https://doi.org/10.1016/j.scitotenv.2023.165865.

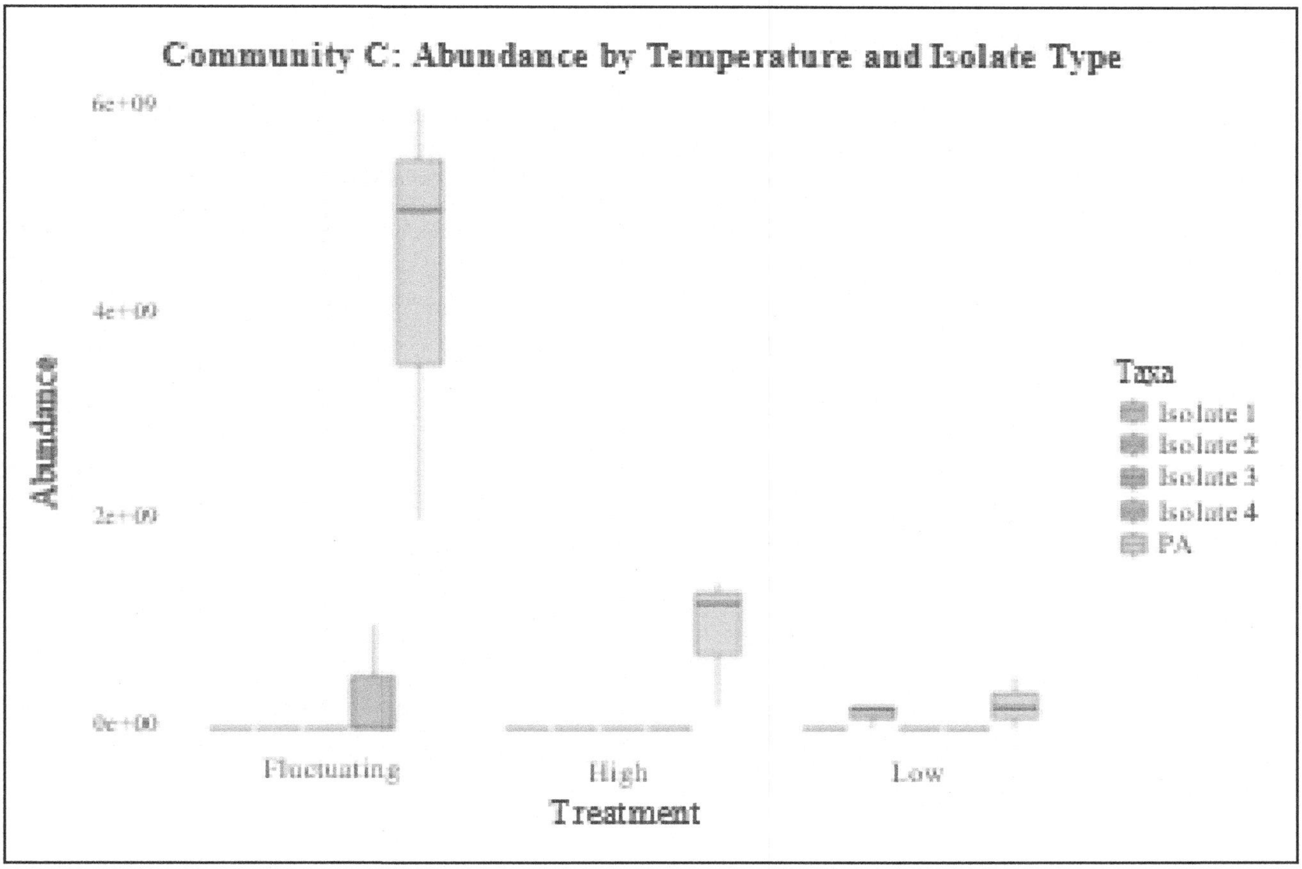

Figure 6. Boxplot illustrating microbial abundance of a synthetic community C composed of four Darlingtonia californica isolates (2, 3, 4, PA) under different temperature treatments. Each color corresponds with a different synthetic community, and the box plots summarize how community-level abundance changes with environmental temperature variability. In this community, PA (P. aeruginosa) had the most dominant abundance under fluctuating and high temperatures, especially in the fluctuating environment, where it reached the highest values. The other isolates have minimal to no growth under all conditions, with only isolate 4 growing slightly under fluctuating conditions and isolate 2 under low temperatures.

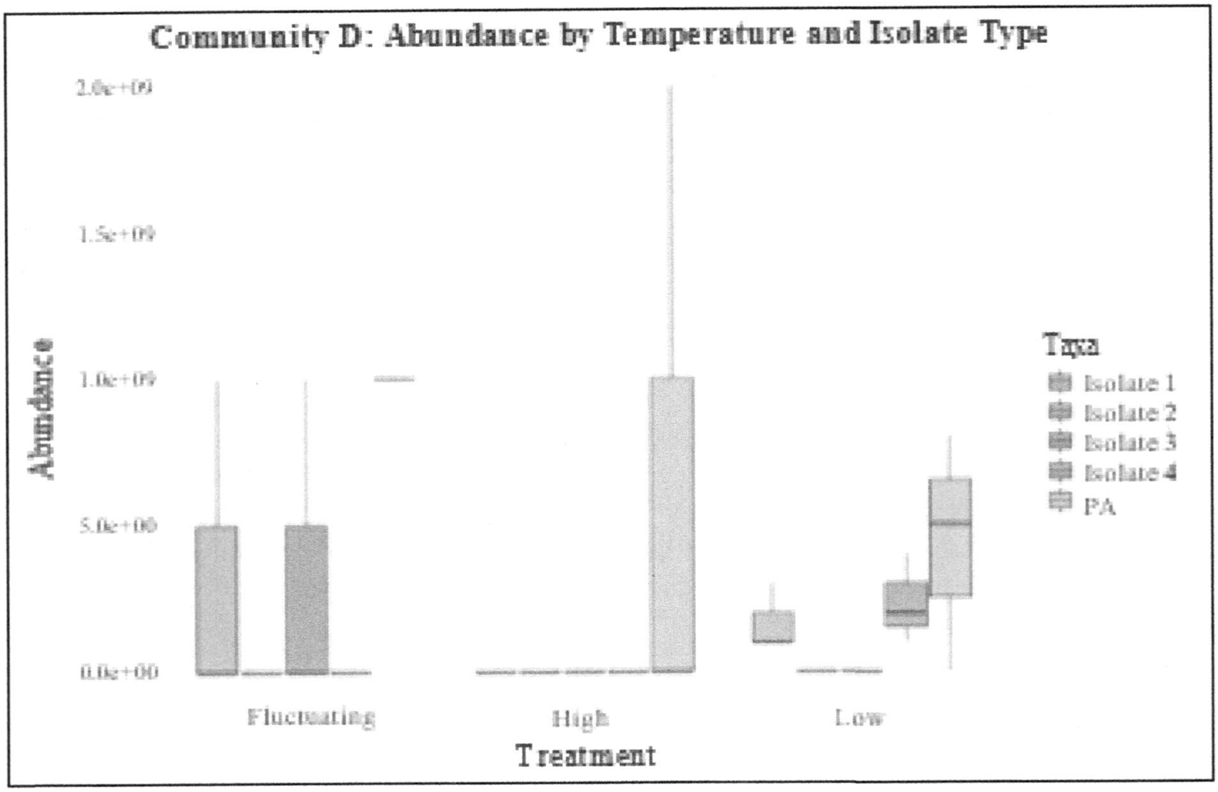

Figure 7. Boxplot illustrating microbial abundance of a synthetic Community D composed of four Darlingtonia californica isolates (1, 2, 3, PA) under different temperature treatments. Each color corresponds with a different synthetic community, and the box plots summarize how community-level abundance changes with environmental temperature variability. PA (P. aeruginosa) showed higher variability and abundance under high temperatures, whereas Isolates 1 and 3 were more active under fluctuating conditions. All isolates showed more even growth under low temperatures.

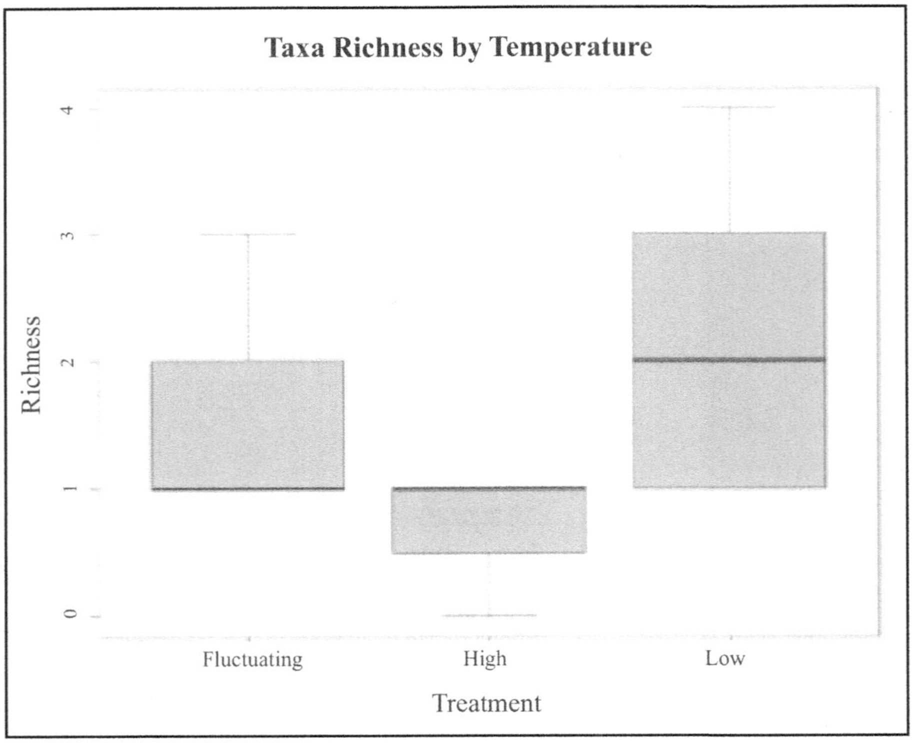

Figure 8. Boxplot showing the number of distinct bacterial taxa (richness) observed across fluctuating, high, and low temperature treatments. Low temperature samples had the greatest richness, followed by fluctuating temperatures. High temperature treatments of communities supported the fewest unique taxa.

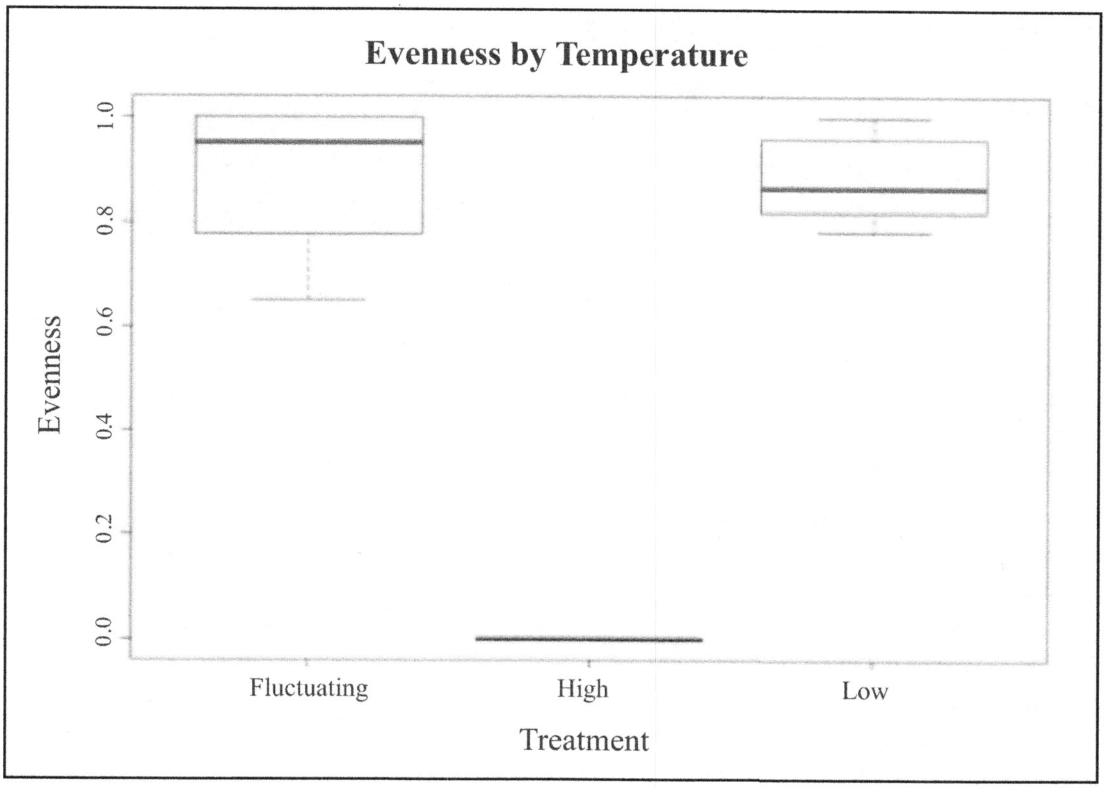

Figure 9. Boxplot showing evenness of microbial communities under different temperature conditions. Evenness was significantly reduced under high temperatures, suggesting dominance by a few species. Both fluctuating and low temperatures promoted more evenly balanced communities.

Humboldt Journal of
Microbiology
2025, Volume 24

Comparative Analysis of Probiotic Viability in Alcoholic and Non-Alcoholic Kombucha

Isabella Cerrone, Chloe Kraft, Brooke Pirkle, Gianna Vendrell, Danielle Williamson

ABSTRACT

Kombucha is a fermented tea beverage valued for its probiotic and antimicrobial properties that arise from complex molecular interactions between yeast, acetic acid bacteria (AAB), lactic acid bacteria (LAB), and metabolic byproducts of fermentation. With the rising demand for non-alcoholic versions of traditionally alcoholic kombucha beverages, ethanol reduction methods have become common, though their impact on probiotic viability and antimicrobial properties remains poorly understood. This study investigates the effects of ethanol removal on kombucha's functional properties, including probiotic survival, antimicrobial activity, and organic acid composition (pH). We compared alcoholic and non-alcoholic versions of JuneShine kombucha across five flavors. Probiotic viability was assessed through colony-forming units (CFUs) of LAB, AAB, and yeasts on selective media plates. Antimicrobial activity against *Escherichia coli* and *Staphylococcus aureus* was measured through agar diffusion assays, and organic acid content was analyzed by pH meter readings. Contrary to our hypothesis, probiotic viability was reduced in alcoholic samples, while no significant changes were observed in antimicrobial efficacy or organic acid profiles between the alcoholic and non-alcoholic kombucha variants. These results suggest a need to reevaluate the health claims of alcoholic kombucha and emphasize the importance of scientific validation in functional beverages.

INTRODUCTION

As consumer demand for non-alcoholic beverages rises, non-alcoholic kombucha production has grown significantly. The global kombucha market, estimated at USD 4.26 billion in 2024, is expected to grow at a compounded annual growth rate (CAGR) of 13.5% from 2025 to 2030 (7). This growth is driving the development of methods to reduce or eliminate ethanol from traditionally brewed kombucha. Commercial non-alcoholic kombucha is typically produced by either controlling fermentation to limit ethanol formation or removing ethanol post-fermentation (13). Many brands promote non-alcoholic kombucha as retaining the same probiotic and antimicrobial benefits as traditional alcoholic kombucha. However, scientific support for these claims remains limited. Ethanol removal could alter microbial viability, organic acid composition, and overall bioactivity, potentially impacting the beverage's health benefits (3).

Kombucha is a fermented beverage made by fermenting sweetened tea with a symbiotic culture of bacteria and yeast (SCOBY), which produces organic acids, polyphenols, and supports probiotic microorganisms beneficial to gut health (9). The microbial community in kombucha is composed primarily of acetic acid bacteria (AAB), lactic acid bacteria (LAB), and yeast species, such as *Saccharomyces cerevisiae* and *Zygosaccharomyces spp*. (15). Kombucha has been shown to have various health benefits, including immune system support and metabolic benefits through its probiotic content and organic acids (3).

Ethanol, a natural byproduct of fermentation, plays an essential role in kombucha flavor development, microbial dynamics, and preservation. Yeast in the SCOBY metabolizes sugars into ethanol, and as the fermentation progresses, it is converted into acetic acid by acetic acid bacteria (11). The balance between ethanol and acetic acid is vital for kombucha's preservation and the survival of beneficial microbes, as too much ethanol produced by yeast can inhibit the growth of AAB, disrupting the fermentation process (9). Ethanol itself is a volatile compound and alone has been shown to possess antimicrobial properties. The ethanol present in kombucha, even at low concentrations as seen in non-alcoholic kombucha (~0.5%), can assist in preventing the overgrowth of harmful microorganisms, which could disrupt the beneficial microbial community in kombucha as well as inhibit the growth of these microorganisms in the gut (13).

While numerous studies have investigated kombucha's probiotic content and antimicrobial effects, there is limited research comparing the functional properties of alcoholic versus non-alcoholic kombucha. Most studies have focused on the bacterial and fungal composition of kombucha and its probiotic potential (9), but few have explored how ethanol removal might affect these characteristics. This knowledge gap is particularly important as the commercialization of non-alcoholic kombucha continues to rise, highlighting the need for research to verify the health claims made about these products (11).

This study aims to investigate the effects of ethanol removal on probiotic viability, antimicrobial activity, and organic acid profiles in kombucha. By comparing alcoholic and non-alcoholic kombucha, we will assess the viability of probiotics, antimicrobial effectiveness, and organic acid concentrations. Probiotic viability will be evaluated using colony-forming unit (CFU) counts, antimicrobial activity will be assessed through diffusion assays against E. coli and S. aureus, and organic acid profiles will be measured through pH meter measurements. This research will provide valuable insights into whether ethanol removal affects kombucha's health benefits and contribute to a better understanding of the differences between alcoholic and non-alcoholic kombucha.

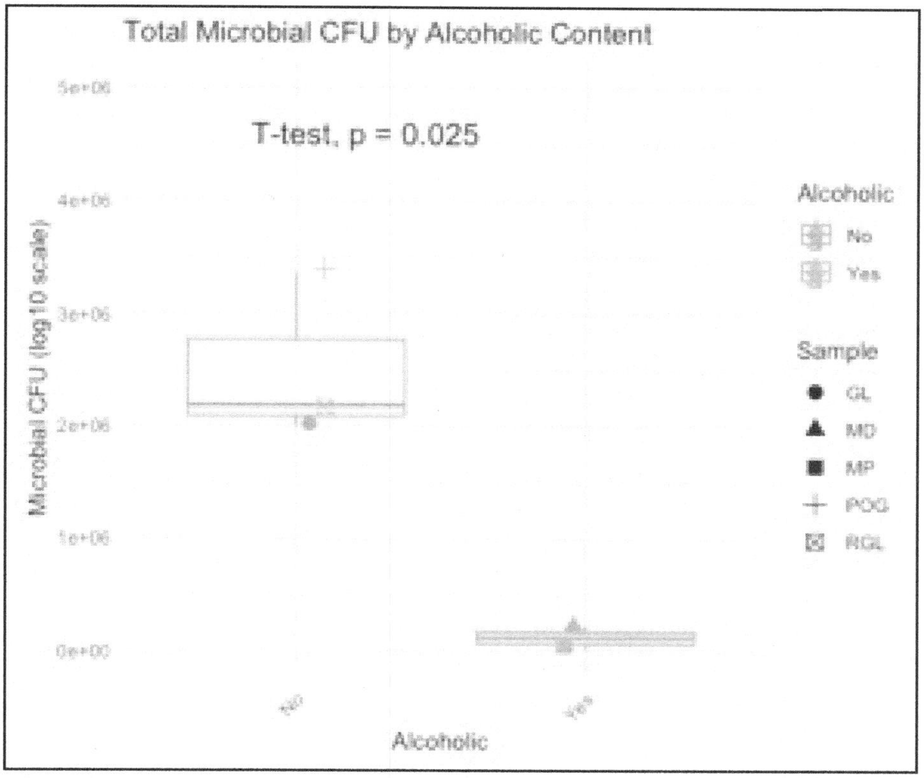

Figure 1. Total microbial counts

This study investigates how ethanol removal affects the functional properties of kombucha, with a focus on bacterial viability, antimicrobial activity, and organic acid composition. We hypothesize that ethanol removal will 1) reduce bacterial and yeast viability by disrupting the balance of microbial communities critical for sustaining beneficial bacteria, 2) decrease antimicrobial activity due to the loss of ethanol's inhibitory effects on pathogenic bacteria, and 3) alter organic acid profiles, resulting in measurable changes in acidity and fermentation metabolites.

To test this hypothesis, we compared alcoholic and non-alcoholic kombucha across five JuneShine flavors using microbiological and biochemical analyses. Probiotic viability was assessed by quantifying colony-forming units (CFUs) of lactic acid bacteria, acetic acid bacteria, and yeasts using selective media. Antimicrobial activity was evaluated through agar diffusion assays against *Escherichia coli* and *Staphylococcus aureus*, and organic acid composition was inferred from pH measurements. Comparative statistical tests were used to determine the significance of differences between the two kombucha types. This research aims to provide insight into whether ethanol removal comprises kombucha's beneficial properties and challenges the viability of health claims made by commercial non-alcoholic kombucha brands.

METHODS

A total of five commercially available kombucha products from the same brand were selected for analysis, the selection of these samples was based on the availability and the categorization of the products as alcoholic (greater than 0.5% alcohol by volume) and non-alcoholic (less than 0.5% alcohol by volume), as specified by the manufactures. The alcoholic flavors were Mango Daydream (MD) and Midnight Painkiller (MP), while the non-alcoholic varieties were Razzberry Ginger Lime (RGL), Ginger Lime (GL), and Passion Fruit Orange Guava (POG). The samples were collected in sterile containers and stored at 4°C before microbiological analysis to minimize microbial degradation.

Microbial quantification was performed by measuring colony-forming units (CFU) for three major microbial groups: yeast, lactic acid bacteria (LAB), and acetic acid bacteria (AAB). Yeast enumeration was conducted by plating serial dilutions of kombucha on selective yeast media (Sabouraud Dextrose Agar, 37°C for 48 hours). Lactic acid bacteria were enumerated using

MRS agar, incubated at 37°C for 48 hours, while acetic acid bacteria were cultured on EMB again, incubated at 37°C for 48 hours. Colony counts were recorded, and the microbial loads for each group were expressed as CFU per millimeter of kombucha.

The pH of each kombucha sample was measured in triplicate using a calibrated pH meter (Acumen AE150, Fisher Scientific). Calibration was performed using standard pH buffers (pH 4.0, 7.0, and 10.0) prior to measurement. The pH was recorded for each sample immediately following the collection.

The antimicrobial properties of kombucha samples were evaluated against *Staphylococcus aureus* and *Escherichia coli* using the Kirby-Bauer disk diffusion method. Inoculations were performed on Muller-Hinton agar, and sterile filter paper discs were impregnated with 10 µL of kombucha. Plates were incubated at 37°C for 24 hours. Zones of inhibition were measured in millimeters (mm), and antimicrobial activity was compared between alcoholic and non-alcoholic samples.

The MIC of the kombucha against *Staphylococcus aureus* and *Escherichia coli* was assessed using the broth microdilution method, following the guidelines of the Clinical and Laboratory Standards Institute (CLSI). Bacterial cultures were grown to mid-log phase in tryptic soy broth, then diluted to a standardized inoculum concentration (approximately 10^6 CFU/mL). Kombucha samples were serially diluted in Eppendorf tubes using saline. An equal volume of diluted bacterial suspension was added to each tube, and the tubes were incubated at 37°C for 24 hours. Tubes were then examined visually for turbidity, with the MIC defined as the lowest kombucha dilution at which noticeable growth occurred.

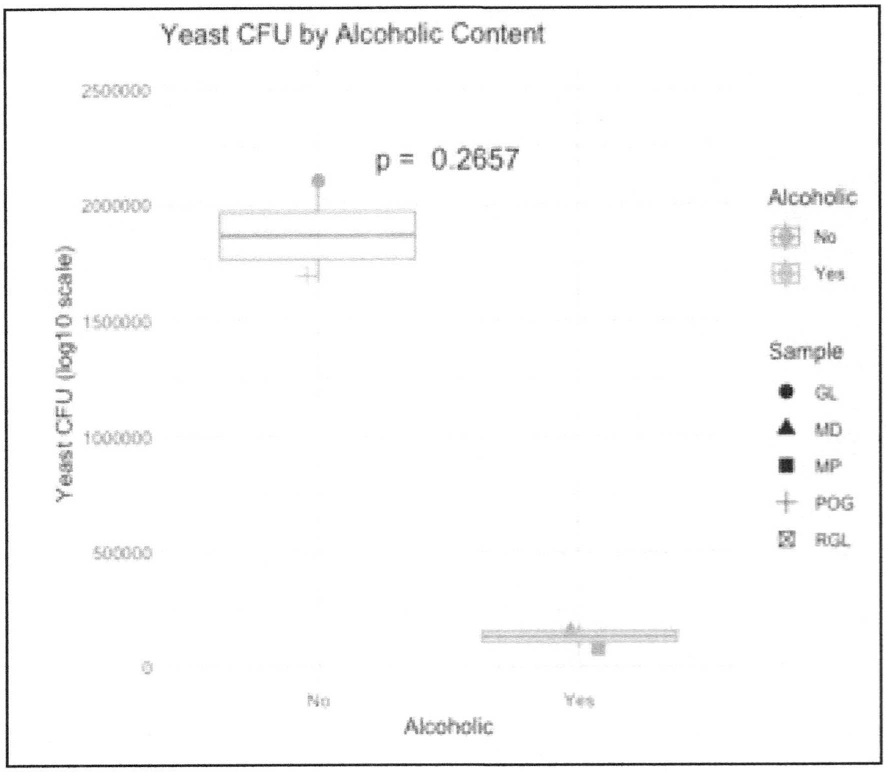

Figure 2. Total yeast counts (CFU/mL) in alcoholic versus non-alcoholic kombucha

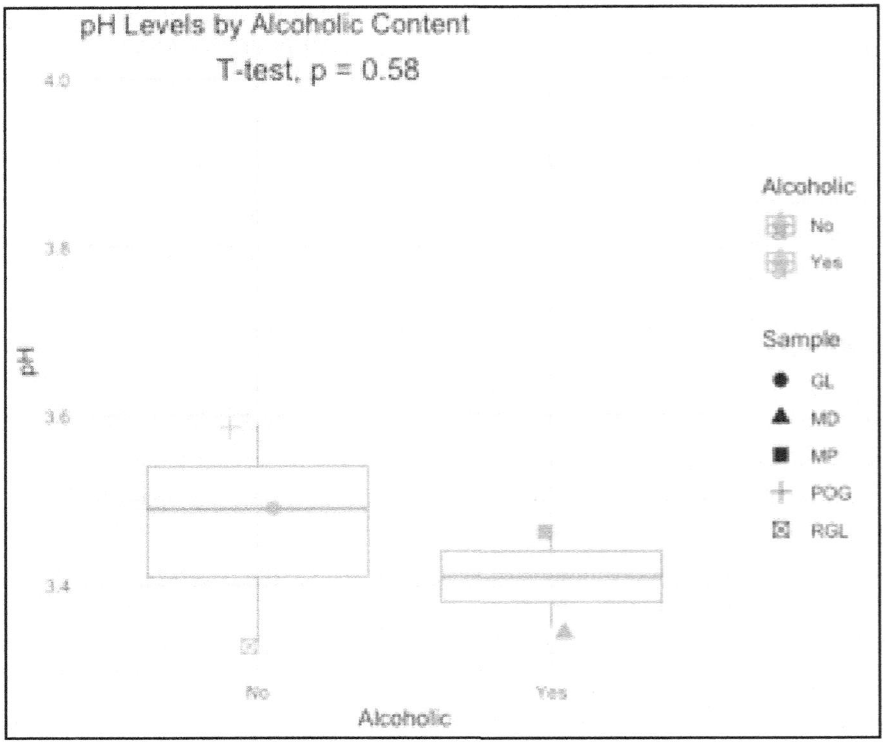

Figure 3. pH levels in alcoholic versus non-alcoholic kombucha samples.

RESULTS

The analysis of microbial populations revealed significant differences in the abundance of bacterial colony-forming unit

(CFU) counts, including combined lactic acid bacteria (LAB) and acetic acid bacteria (AAB) CFUs, between alcoholic and non-alcoholic kombucha (Fig. 1). A two-sample t-test revealed this difference to be statistically significant, $t(8) = 2.774$, $p = 0.02296$, demonstrating that the presence of alcohol markedly reduced probiotic bacterial counts. Non-alcoholic kombucha supported substantially higher bacterial growth compared to alcoholic kombucha.

Yeast CFU counts differed between alcoholic and non-alcoholic kombucha (Fig. 2). The pH values of alcoholic and non-alcoholic kombucha samples were comparable. A two-sample t-test showed no significant difference, $t(3) = -0.56$, $p = 0.6157$, suggesting that alcohol content did not notably influence the acidity of the kombucha samples; however, an independent two-sample t-test showed the difference was not statistically significant, $t(3) = -1.36$, $p = 0.2657$. While the mean yeast CFU count was higher in non-alcoholic kombucha, and visual differences were apparent, high variability and a small sample size likely contributed to the lack of statistical significance.

The pH values of alcoholic and non-alcoholic kombucha samples were comparable (Fig. 3). A two-sample t-test showed no significant difference, $t(3) = -0.56$, $p = 0.6157$, suggesting that alcohol content did not notably influence the acidity of the kombucha.

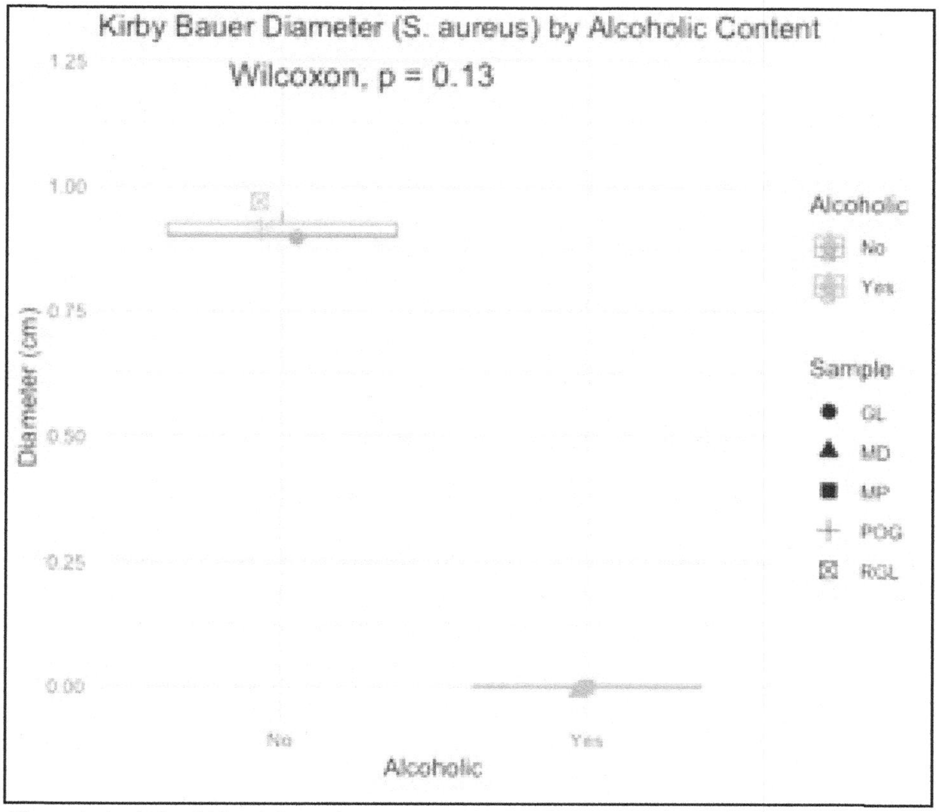

Figure 4. Antimicrobial activity against

Regarding antimicrobial activity, Kirby-Bauer zone of inhibition tests showed that neither alcoholic nor non-alcoholic kombucha inhibited *Escherichia coli* growth. No inhibition zones were observed for any samples. A Wilcoxon rank sum test with continuity correction was used to compare the Kirby Bauer zone of inhibition measurements for *Staphylococcus aureus* (Fig. 4). The test showed no statistically significant difference between alcoholic and non-alcoholic kombucha ($W = 0$, $p = 0.1281$), indicating insufficient evidence for a difference in median antimicrobial activity. Notably, no inhibition zone was observed for *S. aureus* exposed to the alcoholic kombucha, suggesting it lacked antimicrobial effectiveness under the conditions tested.

Minimum inhibitory concentration (MIC) testing further supported these findings (Fig. 5). No inhibition of *Escherichia coli* growth was observed from either alcoholic or non-alcoholic kombucha type, even at the highest concentrations tested. For *S. aureus*, a Wilcoxon rank sum test comparing the minimum inhibitory concentrations (MIC) values between alcoholic and non-alcoholic kombucha showed no significant difference ($W = 1.5$, $p = 1$). Importantly, no MIC could be determined for the alcoholic kombucha (MD), indicating that it failed to inhibit S. aureus growth at any concentration tested.

DISCUSSION

Non-alcoholic kombuchas had significantly higher colony-forming units (CFUs) for lactic acid (LAD) and acetic acid bacteria (AAB) compared to alcoholic kombucha, suggesting an inhibitory effect of ethanol on bacteria viability. Ethanol is known to exert antimicrobial effects by disrupting lipid membranes and denaturing proteins, thereby compromising bacterial cell integrity (14). Contrary to our initial hypothesis that ethanol removal would reduce microbial viability, the absence of alcohol appeared to support more robust microbial populations. These results challenge claims commonly made by kombucha producers that both alcoholic and non-alcoholic kombucha

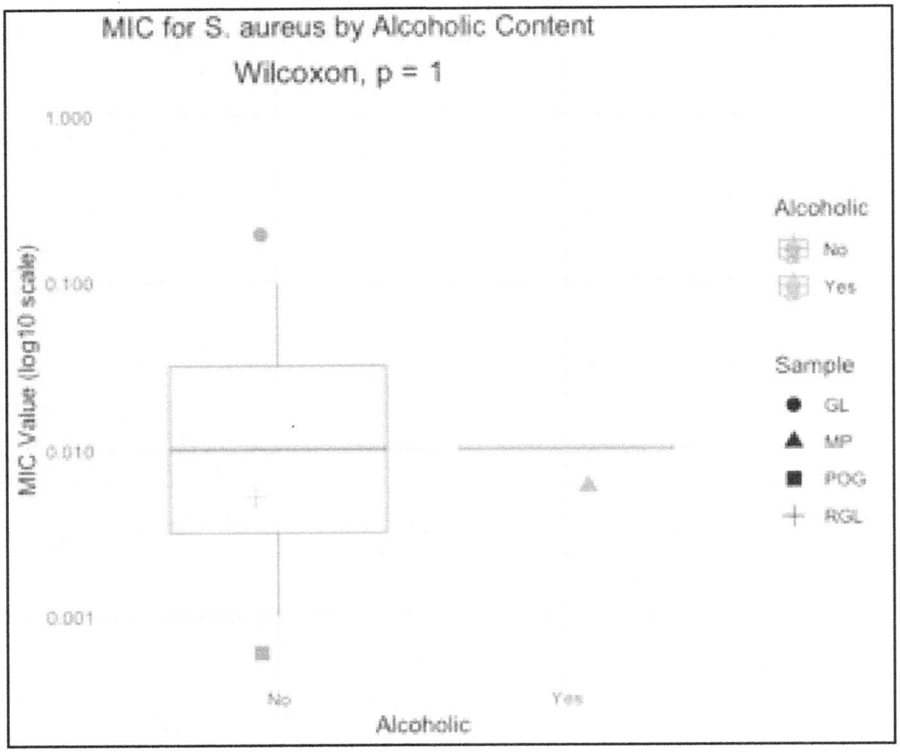

Figure 5. Minimum inhibitory concentration (MIC) against Staphylococcus aureus in alcoholic versus non-alcoholic kombucha

Although previous studies suggest that environmental pH, microbial viability, and interactions (12), we observed no significant pH differences between alcoholic and non-alcoholic kombucha samples. This finding suggests that pH is not a confounding variable in the observed differences in microbial abundance. Our original hypothesis posited that lower pH in alcoholic kombucha might contribute to suppressed microbial growth, but our data did not support this. Minor pH variation may have been driven by various favoring agents, such as lemon or pineapple, rather than fermentation or alcohol content. For a more mechanistic understanding, future work should asses the growth optima of individual microbial isolates across a range of varied pH conditions.

Claims that kombucha inhibits

formulations provide equivalent probiotic benefits. To further evaluate these assertions, future studies should include species-level identification of microbes and examine their functional roles in gastrointestinal colonization, particularly under varying ethanol concentrations.

Yeast quantification did not yield statistically significant differences between alcoholic and non-alcoholic kombucha (Figure 2), although non-alcoholic samples visually exhibited higher yeast CFUs. This lack of significance may be attributed to the small sample size (n = 2 alcoholic, n = 3 non-alcoholic flavors). Notably, one alcoholic sample, Midnight Painiller, yielded no yeast growth, whereas Mango Daydream produced small numbers of colonies. These differences may reflect variation in yeast species during fermentation, which can be influenced by ethanol tolerance. For example, some commercial fermentations may not selectively enrich ethanol-tolerant yeasts in alcoholic kombucha samples (8). Additionally, Midnight Painkiller's inclusion of charcoal, a known adsorbent, could have inhibited yeast growth (6). These findings underscore the importance of controlling for formulation additives and standardizing flavor types in future experiments. Increased replication and inclusion of unflavored kombucha would improve the resolution of microbial population dynamics across product types.

pathogenic bacteria, Escherichia coli and Staphylococcus aureus, in the human gastrointestinal (GI) tract (4) were not strongly supported by our antimicrobial assays. While Kirby-Bauer and minimum inhibitory concentration (MIC) tests showed potential inhibition of Staphylococcus aureus by non-alcoholic kombucha, no inhibition was observed against Escherichia coli. Interestingly, Kirby-Bauer plates exhibited growth of morphologically distinct colonies on kombucha-impregnated filter discs, suggesting that microbes inherent to kombucha may proliferate and compete with the test organisms through microbial interference rather than direct antagonism. It is also possible that ethanol negatively impacted the bioavailability or potency of antimicrobial compounds generated during fermentation.

Several factors may explain the overall limited antimicrobial activity observed: 1) the specific microbial strains used in this study may differ from strains naturally colonizing the human GI tract, 2) fermentation duration and microbial consortia vary between products and could influence metabolite profiles, and 3) our sample size and methodological variability may have reduced statistical power. Future investigations should incorporate well-characterized probiotic and pathogenic strains, larger sample sizes, and metabolic profiling to determine the antimicrobial efficacy of kombucha better.

LITERATURE CITED

1. Bauer-Petrovska, B., & Petrushevska-Tozi, L. (2000). Mineral and water soluble vitamin content in the Kombucha drink. International Journal of Food Science & Technology, 35(2), 201–205. https://doi.org/10.1046/j.1365-2621.2000.00342.x
2. Batista, P., Penas, M. R., Pintado, M., & Oliveira-Silva, P. (2022). Kombucha: Perceptions and Future Prospects. Foods, 11(13), 1977. https://doi.org/10.3390/foods11131977
3. Bazzicalupo, A., & Ruzzi, M. (2020). The Role of Fermentation in the Production of Functional Beverages. Food Research International, 134, 109116.
4. Elzain, A. M., Elsanousi, S. M., & Ibrahim, M. E. A. (2019). Effectiveness of ethanol and methanol alcohols on different isolates of Staphylococcus species. Journal of Bacteriology & Mycology Open Access, 7(4). https://doi.org/10.15406/jbmoa.2019.07.00247
5. Kaewkod, T., Bovonsombut, S., & Tragoolpua, Y. (2019). Antimicrobial activity of kombucha fermented beverage (KFB) singly and in combination with medicinal plants against pathogenic bacteria. ScienceAsia, 45(6), 392-400.
6. Kim, S.-K., Park, D.-H., Song, S. H., Wee, Y.-J., & Jeong, G.-T. (2013). Effect of fermentation inhibitors in the presence and absence of activated charcoal on the growth of Saccharomyces cerevisiae. Bioprocess and Biosystems Engineering, 36(5), 659–666. https://doi.org/10.1007/s00449-013-0888-4
7. Kombucha market size and share: Industry Report, (2030). Kombucha Market Size And Share | Industry Report, 2030. Retrieved from https://www.grandviewresearch.com/industry-analysis/kombucha-market
8. Liu, Y., Zhang, Y., & Zhang, Y. (2021). Isolation, identification, and kinetic and thermodynamic characterization of ethanol-tolerant yeast strains for potential use in kombucha fermentation. Process Biochemistry, 104, 1-10.
9. Marsh, A. J., O'Sullivan, O., Hill, C., Ross, R. P., & Cotter, P. D. (2014). Sequence-based Analysis of the Bacterial and Fungal Compositions of Multiple Kombucha (Tea Fungus) Samples. Food Microbiology, 38, 171-178.
10. McDonald, L. C., & O'Neill, C. E. (2017). Ethanol and Its Impact on the Growth and Viability of Bacterial Communities in Fermented Foods. Journal of Applied Microbiology, 122(2), 347-360.
11. McKenna, K. A., Vickers, A. E., & Liddy, H. L. (2016). The Impact of Ethanol on the Probiotic Viability of Fermented Beverages: A Comparative Study of Alcoholic and Non-Alcoholic Kombucha. International Journal of Food Science and Technology, 51(9), 1801-1809.
12. Ratzke, C., & Gore, J. (2018). Modifying and reacting to the environmental pH can drive bacterial interactions. PLoS Biology, 16(3), e2004248. https://doi.org/10.1371/journal.pbio.2004248
13. Sharma, M., Kumar, S., & Choudhary, S. (2019). Non-Alcoholic Kombucha: Methods for Alcohol Removal and Impact on Functional Properties. Food Research International, 113, 205-211.
14. Valle, D. L., Cabrera, E. C., Puzon, J. J. M., & Rivera, W. L. (2016). Antimicrobial Activities of Methanol, Ethanol, and Supercritical CO2 Extracts of Philippine Piper betle L. on Clinical Isolates of Gram Positive and Gram Negative Bacteria with Transferable Multiple Drug Resistance. PLoS ONE, 11(1), e0146349. https://doi.org/10.1371/journal.pone.0146349
15. Yang, J., Lagishetty, V., Kurnia, P., Henning, S. M., Ahdoot, A. I., & Jacobs, J. P. (2022). Microbial and chemical profiles of commercial kombucha products. Nutrients, 14(3), 670. https://doi.org/10.3390/nu14030670

Humboldt Journal of
Microbiology
2025, Volume 24

Solvent extracts of *Pulvigera papillosa* (Orthotrichaceae) show no antimicrobial properties in disk diffusion tests

Justin Paulin, Michael Lanier, Ryan Solorzano, Grayson Prater

ABSTRACT

As antibiotic-resistant bacteria become more prevalent, the search for novel forms of antibiotics in the natural world has become an increasingly worthwhile endeavor. Many mosses have been found to contain antimicrobial properties, including those in the family Orthotrichaceae. *Pulvigera papillosa*, an Orthotrichaceous moss native to California, has thus far not been investigated to determine whether it contains antimicrobial properties. Extraction techniques using polar and nonpolar solvents have proven to be effective in extracting the secondary metabolites that allow mosses to defend themselves against a hostile microbial world. This study takes a preliminary look into the antimicrobial properties of *P. papillosa* using methanol and hexane extractions in conjunction with Kirby-Bauer disk diffusion tests. Evidence for the presence of antimicrobial compounds within the tissues of *P. papillosa* was not found using our methods. However, our study does not conclusively eliminate *P. papillosa* as a potential source of antimicrobial compounds.

INTRODUCTION

The advent of modern antibiotics ushered in an era where common infections are no longer a death sentence. Beginning with Alexander Fleming's discovery of Penicillin in 1928, the use of antibiotics in treating infectious diseases became increasingly prevalent, resulting in a significant increase in average life expectancy. From 1900 to 2021, the average life expectancy of adults in the United States increased approximately three decades (1). This significant advancement in medicine made infections that were once a death sentence eminently survivable. Despite the enormously increased ability of medical professionals to treat infectious diseases, the use of modern antibiotics has not been a silver bullet. Bacterial strains often quickly become resistant to antibiotics due to their rapid reproduction and horizontal transfer of genetic material (2). Antibiotic-resistant bacteria emerged early on in the use of modern antibiotics and, over time, resistance has only become more widespread. This has led to a dynamic where new antibiotics must be rapidly developed to keep pace with the development of antibiotic resistance in bacteria.

Figure 1. *Pulvigera papillosa*

The use of natural products, such as moss, as sources of antibiotic discoveries is pervasive. In fact, ~75% of antibiotics used today are derived from natural products (3). Mosses are a clade of nonvascular plants comprising approximately 14,000 species. They lack physical barriers to the external environment beyond a thin cuticle, making them susceptible to invasion from microorganisms. Many moss species thrive in dim, humid, environments that are well-suited to bacterial growth. However, mosses are rarely the target of herbivores and pathogenic microbes, in contrast to their vascular counterparts (4). This may be due to the many secondary metabolites (benzenoids, bibenzyls, fatty acid derivatives, flavonoids, phenylpropanoids, terpenes, and terpenoids) found within moss tissue which may act as a deterrent to herbivory and pathogenic activity. Some of these compounds have been isolated and shown to have antimicrobial properties (5). *Orthotrichum rupestre* (Orthotrichaceae) was one of the first mosses to be investigated for antimicrobial properties (6). Results from this experiment by McCleary et al. showed that extracts from this species inhibited bacterial growth (7). Given that biochemistry is frequently conserved among related taxa, the confamilial California-native moss *P. papillosa* is certainly worth investigating for these properties as well.

Antibiotic resistance in bacteria is commonly studied in the use of ESKAPE bacteria. ESKAPE is an acronym for 6 or more model bacteria species that include *Escherichia coli/Enterococcus faecium*, *Staphylococcus aureus*, *Klebsiella pneumoniae*, *Acinetobacter baumannii*, *Pseudomonas aeruginosa*, and *Enterobacter* spp. This group comprises some of the most virulent pathogenic bacteria. Many potentially fatal infections arise from exposure to these microorganisms. ESKAPE bacteria are responsible for 80% of bacteria-related deaths worldwide and the expected global death toll due to antibiotic-resistant bacteria is ten million by the year 2050 (8). Additionally, because ESKAPE bacteria are known to develop resistance to multiple drugs, studying how to combat them is critically important for public health.

In this study, we explored the possible antimicrobial properties of the native moss *P. papillosa*. To assess the possible antimicrobial properties, we washed the moss in a 1% bleach solution and hoped to extract secondary metabolites from the moss using polar dominant and non-polar solvents. We then impregnated Kirby-Bauer disks with our extracts and observed zones of inhibited growth within cultures of ESKAPE bacteria. Zones of inhibition around the moss extract treated disks would be indicative that the moss holds antibacterial properties. Zones of inhibition were expected to be measured and compared with each other and controls, allowing us a preliminary look at this moss's antimicrobial properties.

MATERIALS AND METHODS

Identification

Pulvigera papillosa was collected from the bark of an ornamental *Acer sp.* (Figures 1 and 2). located near the Cal Poly Humboldt library. The moss was identified to species by the first author, using an Amscope B120 light microscope and Nickte Mendez's "Key to Orthotrichaceae in California" (https://paulsiriwilson.space/orthotrichum/).

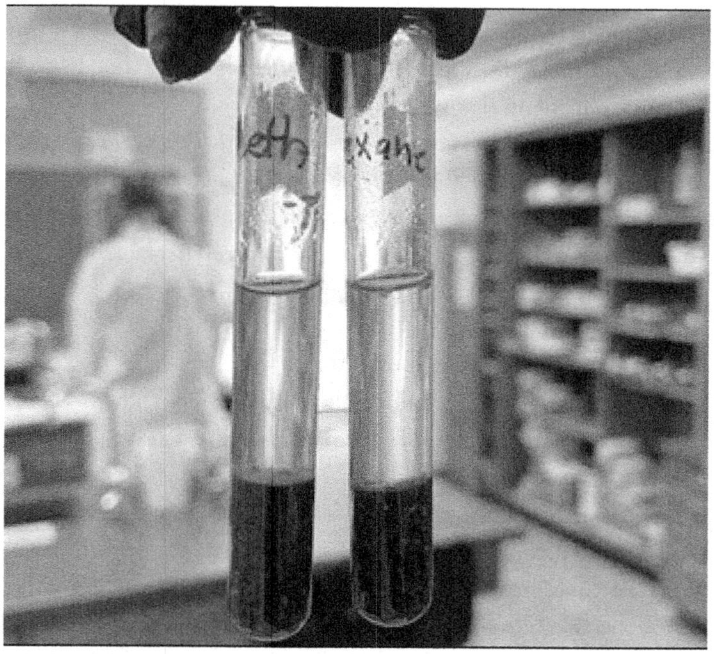
Figure 2. Unconcentrated methanol (left) and hexane (right) moss extracts

Pulverization of Moss and Extraction of Metabolites

5.7022 grams of *P. papillosa* was collected from the same tree from which it was identified. The moss was first sterilized in a 1% bleach solution, then left to soak in the solution for 10 minutes before being rinsed with sterile water to remove excess bleach.

Sterilized moss was placed in a mortar and liquid nitrogen was added. The moss was then pulverized with a pestle until finely crushed. The pulverized moss was freeze dried for 24 hours to remove remaining liquid.

Two batches of extraction mixtures were assembled, one using a polar solvent and one using a non-polar solvent, in order to extract both polar and non-polar metabolites (Figure 4). Batch A contained 0.5 grams of pulverized moss and 9 mL of 100% methanol (a polar solvent). Batch B contained 0.5 grams of pulverized moss and 9 mL of hexane (a nonpolar solvent). Both batches were left to sit for 48 hours and were agitated periodically. Three additional solutions were used as control groups; 100% methanol and 100% hexane were used as negative controls while Kanamycin, an aminoglycoside antibiotic, was used as a positive control.

Kirby-Bauer Disk Impregnation and Treatment Plating

The supernatant from each extraction mixture was removed and filtered by a syringe with a 0.22 μm syringe filter. Ten μL of liquid was pipetted onto each Kirby-Bauer disk.

Each treatment (Batch A methanol extracts, control of 100% methanol, Batch B hexane moss extracts, control of

100% hexane, antibiotic positive control species; Kanamycin) was applied to 15 disks. All disks were dried in a fume hood before placement on agar plates.

ESKAPE bacteria species (*Acinetobacter* sp. *Pseudomonas aeruginosa, Enterobacter faecalis*) were spread plated onto TSA plates. Each bacterial species was plated on five plates. A Kirby-Bauer disk of each treatment type was placed on every plate. All plates were left to incubate at 37°C for 24 hours.

Concentration of moss extracts.

Methanol and hexane are highly volatile compounds. In order to concentrate solutions, both Batch A and Batch B were transferred into sterile tubes and allowed to evaporate for 5 days. Concentrated batch extracts were dissolved in 400ul of solvent to allow for pipetting. Kirby Bauer disks were impregnated with 10 μL of concentrated batch extracts. Plates were inoculated with respective ESKAPE bacteria and treated with the newly concentrated Batch A and B extracts, as well as the previous control solutions.

Additional tests.

Six plates were inoculated with ESKAPE bacteria, two plates for each bacterial species. One set of plates was treated with pulverized moss without sterilization. This was done to test the interaction between pathogenic bacteria and ecological bacteria found on the surface of the moss.

A set of bacteria-inoculated plates was treated with freshly collected unaltered moss to observe possible bacterial competition between moss bacteria and the ESKAPE bacteria. Additionally, intact moss was soaked in a beaker of 30 mL of deionized water. A sample from the water was used to streak a separate plate to observe the growth of bacteria that resides on the surface of the moss. All results were documented with an OM System OM1 mirrorless camera.

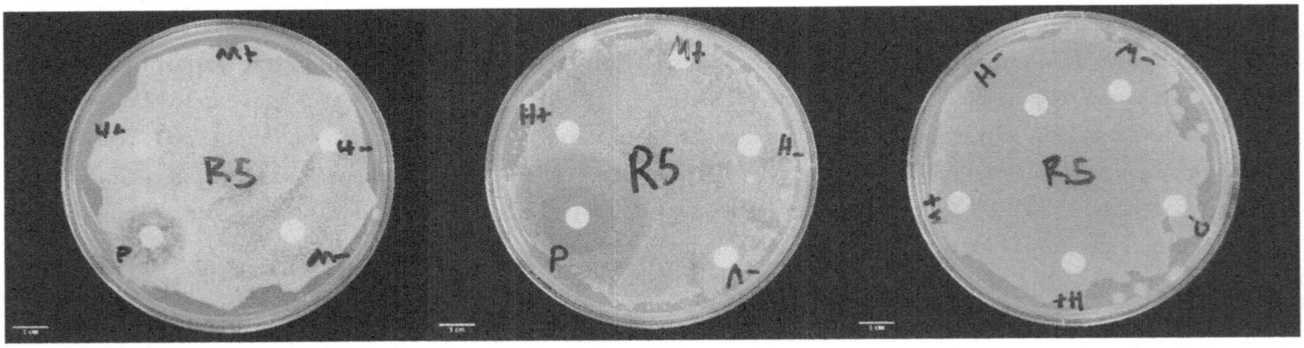

Figure 3. Bacterial cultures of *P. aeruginosa* (right), *Acinetobacter sp.* (left), and *E. faecalis* (center) with Kirby-Bauer disks. Only Kanamycin-impregnated Kirby-Bauer disks (P) caused inhibition in the middle and bottom plates).

Table 1. Zone of inhibition diameters for Kanamycin control group treatments were averaged, other treatment measurements presented on a +/- basis

Species	Kanamycin	Batch A	Batch B	100% Methanol	100% Hexane
Acinetobacter baumannii	1.65 cm	-	-	-	-
Enterococcus faecalis	3.84 cm	-	-	-	-
Pseudomonas aeruginosa	0.00 cm	-	-	-	-

RESULTS

Kirby-Bauer disks impregnated with Batch A extract, Batch B extract, 100% methanol, and 100% hexane produced no zones of inhibition (Figure 3). This was the case for every bacterial species tested. Kanamycin-impregnated disks did produce zones of inhibition in every tested bacterial species except for *Pseudomonas aeruginosa*. Table 1 lists all results for the Kirby-Bauer disk diffusion tests.

The three TSA agar plates grown from a wash of live moss grew countless colony-forming units of a single colonial morphology (Figure 5). This species was not identified definitively, but the rough appearance of the colonies indicates an affinity to *Bacillus*.

DISCUSSION

This study shows that antimicrobial metabolites cannot be extracted from *P. papillosa* with our methods, suggesting that the moss *P. papillosa* may not rely on polar or nonpolar secondary metabolites as antimicrobial agents. Methanol and hexane, the solvents used in this study, have shown to be effective at extracting such compounds in other moss species (9), yet no antimicrobial activity was observed when using our extracts in disk diffusion tests.

While we did not observe antimicrobial activity in any of our experimental treatments, there is a possibility that these results reflect the limitations of our methods rather than a true absence of antimicrobial compounds in the moss. To sanitize the moss, we soaked it in a bleach solution, unlike Bogomolnaya et al., who grew their moss in a sterile environment (9). It's possible that the bleach treatment rendered the moss metabolites inert or that the soaking period leached all antimicrobial compounds from the moss. Additionally, our extracts may have not been concentrated enough to display inhibition. It is possible that with different methods antimicrobial compounds may be extracted from this moss. However, it is fairly conclusive that the small organic compounds likely to be captured by our polar and nonpolar solvents do not have antimicrobial properties.

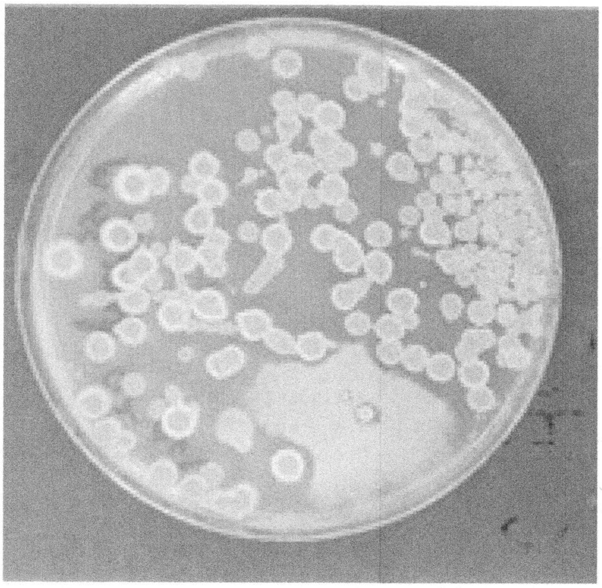

Figure 4. *Pulvigera papillosa* wash and streak plate. This shows a single morphology of bacteria thought to be a *bacillus sp.*

In the absence of antimicrobial secondary metabolites, how does *P. papillosa* protect itself from harmful microbes? Our preliminary exploration into the microbiome of *P. papillosa* may give us a hint into its strategy. The streak plates taken from our moss's surface revealed a single bacterial species dominating the moss. This suggests a potential host-specific interaction, possibly a symbiotic relationship between this moss and this bacterium. This manner of host specific interactions are known with mosses in the genus *Sphagnum*. However, this genus is known to carry a diverse group of symbiotic bacteria. Moss-bacterial associations are known to help other moss species fix carbon or nitrogen, sequester carbon, maintain moisture levels and fend off pathogens (10). It is also possible that *P. papillosa* uses chemical signaling to engineer its microbiome creating favorable conditions for this bacterium, as other mosses and plants have been found to do (11). Further studies might pursue biochemical and metabolic testing of the bacterial species associated with *P. papillosa* to identify it and help determine what functions, if any, it provides for the moss.

While our extractions did not reveal any antimicrobial secondary metabolites in *P. papillosa*, our investigations into this species of moss has led to interesting insights into its ecology. The apparent presence of a single bacterial species dominating the external surfaces of this moss warrants further investigation into P. papillosa's relationship with its epiphytic microbiome.

REFERENCES

1) Arias E, Xu J, Kochaneck K. 2021. United States Life Tables, 2021. Natl Vital Stat Rep 72(12)
2) Dadgostar P. 2019. Antimicrobial Resistance: Implications and Costs. Infect Drug Resist 12, 3903–3910.
3) Aruwa CE, Chellan T, S'thebe NW, Dweba Y, Sabiu S. 2024. ESKAPE pathogens and associated quorum sensing systems: New targets for novel antimicrobials development. Health Science Review. 11:100155
4) Saxena K, Yadav U. 2018. In vitro assessment of antimicrobial activity of aqueous and alcoholic extracts of moss Atrichum undulatum (Hedw.) P. Beauv. Physiol

Mol Biol Plants. 24(6):1203-1208

5) Valeeva LR, Dague AL, Hall MH, Tikhonova AE, Sharipova MR, Valentovic MA, von Reuß SH, König WA. 2005. Olefinic isothiocyanates and iminodithiocarbonates from the liverwort Corsinia coriandrina. Eur. J. Org. Chem. 1184-1188
6) Dziwak M, Wróblewska K, Szumny A, Galek R. 2022. Modern use of bryophytes as a source of secondary metabolites. Agronomy (Basel) 12(6):1456.
7) McCleary JA, Sypherd PS, Walkington DL. 1960. Mosses as possible sources of antibiotics. Science 131(3393):108–108.
8) De Oliveira DMP, Forde BM, Kidd TJ, Harris PNA, Schembri MA, Beatson SA, Paterson DL, Walker MJ. 2020. Antimicrobial Resistance in ESKAPE Pathogens. Clin Microbiol Rev 33:10.1128/cmr.00181-19.
9) Bogomolnaya LM, Shakirov EV. 2022. Antimicrobial activities of secondary metabolites from model mosses. Antibiotics (Basel) 11(8):1004.
10) Baev V, Gecheva G, Apostolova E, Gozmanova M, Yahubyan G. 2024. Exploring the Metatranscriptome of Bacterial Communities of Two Moss Species Thriving in Different Environments-Terrestrial and Aquatic. Plants (Basel) 13(9):1210.
11) Alvarenga DO, Rousk K. 2022. Unraveling host–microbe interactions and ecosystem functions in moss–bacteria symbioses. J Exp Bot 73(13): 4473–4486.

Humboldt Journal of Microbiology
2025, Volume 24

Assessing Antimicrobial Properties of Plant-derived Ellagic Acid Supplements

Noah Schuhmann, Andrew McLaughlin, Julian Barrera

ABSTRACT

Bacteria present in the human gut microbiota are affected by compounds with antimicrobial properties. Phenolic compounds extracted from plants have been shown to exhibit significant antimicrobial capacity at certain concentrations, which makes them an area of interest to researchers. Ellagic acid is one such phenolic compound, present in nuts and berries that many people consume on a regular basis. Supplements containing concentrated ellagic acid provide an additional source for the compound. It is known that, in the body, ellagic acid is metabolized into urolithins, which contribute to health through their anticancer and anti-inflammatory qualities. In this study, we investigated the ability of ellagic acid to inhibit growth of bacteria common in the human gut microbiome and determined the concentrations at which it is most effective in doing so. We found that, despite being polyphenolic, ellagic acid had no visible inhibitory effect on the bacterial species that were tested.

INTRODUCTION

Organic plant extracts, particularly phenolic compounds, have been characterized as potential antimicrobial agents. Phenols are characterized by a hydroxyl (-OH) group attached to the carbon of an aromatic ring. Plant extracts with higher concentrations of phenols more effectively inhibit bacterial growth than those with lower concentrations (1), though the mechanism by which phenols produce these effects is not fully understood. As a result, certain plants are actively explored for their potential to affect microbial growth (2) and are tested against both Gram-positive and negative bacteria (3). Animals that ingest these plants may be affected by these compounds, depending on how certain phenols interact with the microbes that make up their gut microbiome (4).

Ellagic acid is found in the fruits and nuts of several plants common in human diets. Urolithins, which play important roles as antioxidants and in regulating tumor suppression pathways (5), are produced from ellagic acid. The process of converting ellagic acid to urolithins in the human body is mediated by bacteria of the genera Gordonibacter and Ellagibacter, which exist in the digestive tract of humans, specifically, the intestines (6), (7). Other species present in the human gut microbiota, which have not been linked with these processes, are potentially negatively impacted by ellagic acid.

A class of molecules related to ellagic acid, ellagitannins, have been shown to contribute significantly to antimicrobial activity (8). Given its structure as a polyphenol, along with the success of other similar plant-derived phenols in producing antimicrobial effects, it is possible that ellagic acid may exhibit antimicrobial activity. Thus far, ellagic acid has been tested at concentrations of 5mg/mL against S. aureus and E. coli, among other bacteria, and has proven to be ineffective at growth inhibition. However, one specific ellagitannin, epigallocatechin gallate, has shown some signs of antimicrobial activity at this concentration, and significant antimicrobial activity at a concentration of 15 mg/mL (8). While ellagic acid may have been ineffective at concentrations of 5 mg/mL, a more recent study has shown that extracts from the plant Euphorbia helioscopia L. inhibited the growth of two strains of MRSA. These extracts contained ellagic acid at high concentrations, relative to some of the other compounds

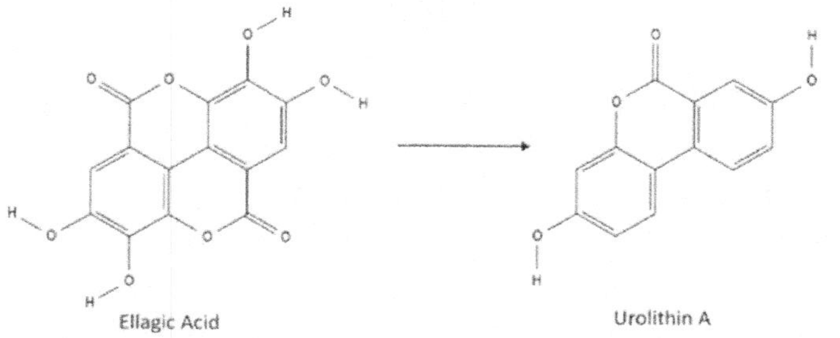

Figure 1. Chemical structures of both ellagic acid and urolithin A. Arrow indicates transformation of ellagic acid into urolithin A via gut bacteria within *Gordonibacter* and *Ellagibacter* genera. Drawn using CAS SciFinder.

present (2). This finding suggests that concentrated ellagic acid may have an inhibitory effect on bacteria within the human gut microbiome. As a dietary component, concentrated ellagic acid supplements are available to consumers.

We are interested in the potential for concentrated ellagic acid found in pomegranate extract supplements to have significant inhibitory effects on bacteria associated with the gut biome, and whether it has different effects based on bacterial cell wall composition. If phenols derived from plant extracts have inhibitory effects against various species of bacteria, then ellagic acid, a polyphenolic compound, will inhibit bacterial growth in the gut biome, and do so independently of Gram-negative or Gram-positive cell walls. To test this, we will utilize the Kirby-Bauer disc diffusion method on separate plated cultures containing different bacterial species present in the human gut microbiota. We expect degrees of inhibition to increase with greater concentration of ellagic acid, and no significant differences in bacterial growth inhibition between Gram-positive and Gram-negative species.

METHODS

All of the bacteria we will use are known to be commonly present in the human gut microbiota. We acquired prepared liquid cultures of our bacteria of interest for this study. They include two Gram-positive bacteria (Enterococcus faecalis and Staphylococcus aureus) and three Gram-negative bacteria (Klebsiella pneumoniae, Proteus vulgaris and Escherichia coli). Each of these bacteria are capable of growth in aerobic conditions.

Excluding E. faecalis, our bacteria were plated using sterile cotton swabs on four Mueller-Hinton (MH) agar plates, which were in turn divided into four regions to implement replication. The culture of E. faecalis was plated on blood-enriched MH agar to provide improved growth conditions, as its growth on standard MH media is suboptimal. We added four 5mm diameter filter paper discs to the quadrants of each plate, one disc per quadrant, and applied the same treatment to three of the discs on one single plate, while the fourth disc received the control treatment.

We prepared three solutions containing different concentrations of ellagic acid supplement acquired from an online vendor, dissolved in 1M NaOH solution. Our three ellagic acid solutions were at concentrations of 50µg/µL, 25µg/µL, and 12.5µg/µL. The most concentrated of these were prepared by dissolving 0.5g of ellagic acid that was collected from five pill capsules in 3.31mL of 1M NaOH solution and 6.69mL of distilled water. The subsequent solutions were prepared from this stock by performing two half dilutions in microcentrifuge tubes. The amount and concentration of hydroxide used were calculated to ensure adequate solubility of our ellagic acid. The solutions were applied to filter paper discs after they were placed by volumetrically pipetting 20µL of the solution onto the appropriate discs. Of the four plates for each bacterium, three received the differing concentrations of ellagic acid supplement solution, and the fourth received the 10µg/disc penicillin G treatment. The control treatment on each plate consists of 20µL of 1M NaOH solution.

Following an incubation period of 48 hours at 37°C, we determined the relative efficacy of each ellagic acid solution to one another and the antibiotic for each bacterium. This was accomplished through direct measurement of the diameter of growth inhibition zones surrounding the discs. Zones of inhibition surrounding our penicillin G discs were also determined and used as a reference with those associated with the ellagic acid treatments.

To determine whether the inhibitory effects of our ellagic acid solutions significantly differed for each species of bacteria, we prepared to utilize a t-test to compare between two experimental means (Figure 2); that being the average zone of inhibition for each treatment. The same test could be used to compare between our different species of bacteria for the same treatment.

A Scanning Electron Microscopy (SEM) micrograph was made to confirm the presence or absence of E. coli growth in the area around one of our diffusion discs where ellagic acid was applied. Agar plugs from media next to diffusion discs were taken from plates that received the highest concentration of ellagic acid solution treatment (50µg/mL) and fixed using 3% glutaraldehyde in a cacodylate buffer for 24 hours, air dried, then sputter-coated with gold, and viewed in SEM.

$$t= |\bar{x}_1-\bar{x}_2| / (s_p \cdot \sqrt{(N_1+N_2)/(N_1 \cdot N_2)})$$

Figure 2. T-test equation for comparing two experimental means. $\bar{x}$ indicates a dataset mean, N indicates the number of replicates within a given dataset, and sp indicates the pooled standard deviation. From *Fundamentals of Analytical Chemistry* (9).

RESULTS

After 48 hours of incubation at 37°C, the average zone of inhibition from the ellagic acid solutions (50µg/mL, 25µg/mL, 12.5µg/mL) was 0.0cm for all bacteria tested (Table 1, Figure 3). Indeed, the cells continued to grow directly on top of the ellagic acid solution (Figure 5). The average zones of inhibition around the 1M NaOH discs were roughly 1.1cm for all bacteria except for *E. faecalis*, which experienced no growth inhibition for this control treatment (Table 1, Figure 3). Both Gram positive bacteria, *S. aureus* and *E. faecalis*, displayed zones of inhibition around the 10µg penicillin G discs, with a greater area exhibited by *S. aureus*. Two of the Gram-negative bacteria, *E. coli* and *K. pneumoniae*, experienced no growth inhibition around the penicillin G treatment discs. However, our third Gram-negative bacteria, *P. vulgaris*, showed an average zone of inhibition around the penicillin G discs comparable to that of *E. faecalis* (Table 1, Figure 4). As our treatments didn't produce zones of inhibition, we did not carry out our planned statistical analyses.

Table 1: Shows all bacteria used, treatments applied to each plate, and resulting average zone of inhibition (cm) and the standard deviation based on three replicates.

Treatment	*S. aureus*	*E. faecalis*	*P. vulgaris*	*E. coli*	*K. pneumoniae*
50 µg/mL	0.0 ± 0.0	0.0 ± 0.0	0.0 ± 0.0	0.0 ± 0.0	0.0 ± 0.0
25 µg/mL	0.0 ± 0.0	0.0 ± 0.0	0.0 ± 0.0	0.0 ± 0.0	0.0 ± 0.0
12.5 µg/mL	0.0 ± 0.0	0.0 ± 0.0	0.0 ± 0.0	0.0 ± 0.0	0.0 ± 0.0
Penicillin G	>3.0	1.6 ± 0.2	1.6 ± 0.1	0.0 ± 0.0	0.0 ± 0.0
Control	1.1 ± 0.1	0.0 ± 0.0	1.0 ± 0.1	1.1 ± 0.1	1.1 ± 0.1

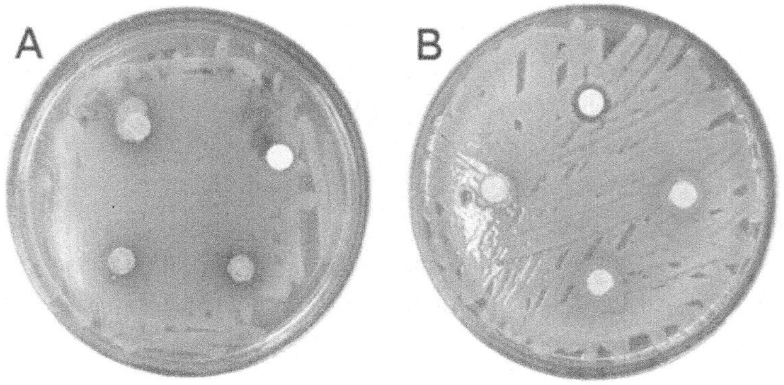

Figure 2. Two MH plates inoculated with bacteria and treated with 50µg/µL ellagic acid solution using diffusion discs containing 20µL of solution, with 1M NaOH as a control. A) P. vulgaris MH plate following treatment using the 50µg/µL ellagic acid solution, resulting in an average zone of inhibition of 0.0cm. B) K. pneumoniae MH plate following treatment using the 50µg/µL ellagic acid solution, resulting in an average zone of inhibition of 0.0cm.

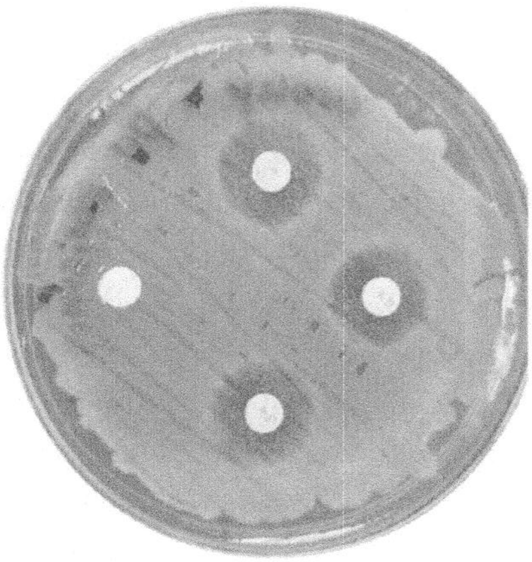

Figure 3. *P. vulgaris* Muller-Hinton plate following treatment using Penicillin G (10µg) diffusion discs, resulting in an average zone of inhibition of 1.6 cm. A blank diffusion disc was also used as a control (left).

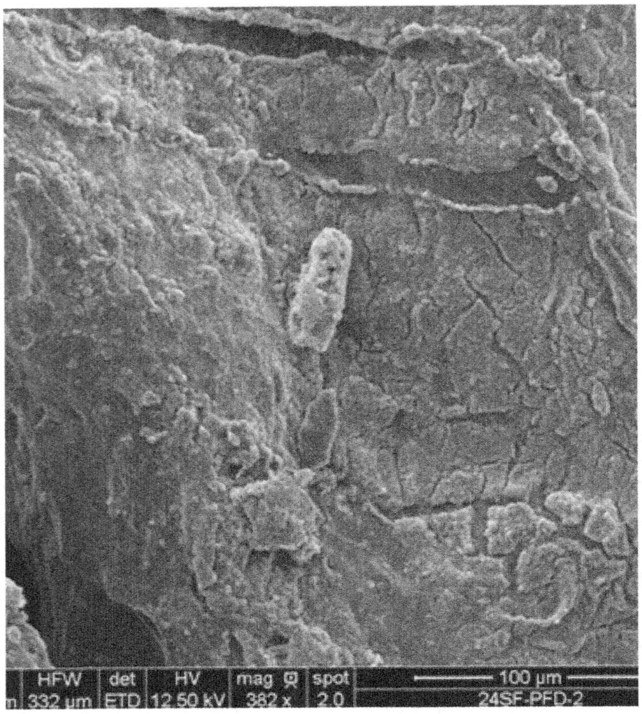

Figure 4. SEM micrograph showing an *E. coli* cell living on media where ellagic acid solution was also present. The "cracks" under the cell and agar surface are dried remnants of ellagic acid solution.

DISCUSSION

There is a large variety of commercially available dietary supplements, many of which claim to produce beneficial health effects for consumers. Such dietary supplements are not regulated, nor are their advertised health effects verified by the Food and Drug Administration (10). Their effects on the human gut microbiota have not been well-characterized. We wanted to determine if concentrated ellagic acid supplements are detrimental to bacteria present in the gut microbiome, despite their positive association with tumor suppression activity.

Our results strongly indicate that ellagic acid was not driving the inhibitory effect observed by Ribinskas and collaborators (2). In that study, a plant extract concentrated in ellagic acid, quercetin, and hyperoside was tested against strains of MRSA, with some strains displaying a zone of inhibition where the extract was applied. This led us to believe that ellagic acid contributed to Ribinskas et al's observed growth inhibition of MRSA and would have similar effects on the gut bacteria we studied.

Certain factors may have influenced our conclusion. We prepared our ellagic acid solutions by dissolving the contents of a supplement containing some quantity of rice powder, silica, and magnesium stearate, and we do not know exactly how this affected the growth of our bacteria. While we approximated the mass of ellagic acid within the capsules, the labeled supplement facts were unclear about the precise content and instead provided a general estimate from which we devised our dilution scheme. We encountered another limitation when deciding on a solvent for the ellagic acid. In a related study, an ellagic acid solution was reportedly prepared at a concentration of 5mg/mL using distilled water as a solvent (8). We found information from chemical data sheets available on the National Institutes of Health website suggesting that ellagic acid is sparingly soluble in water, at around 0.8mg/mL. This discrepancy led us to favor the NaOH solution as a solvent to ensure dissolution of the ellagic acid at our desired concentrations. The growth conditions for our bacteria were aerobic, which does not quite align with the microaerophilic environment of the human gut. Closely matching their laboratory growth conditions to those found in the gut would have improved consistency with our hypothesis.

Though we found it to have no antimicrobial properties, ellagic acid may affect bacterial growth in some other way that could not be determined in our study. It is possible that ellagic acid does contribute to bacterial growth inhibition, but only in conjunction with other phenolic compounds like those present in the plant extract of the previously mentioned study. Another consideration is that the presence of ellagic acid - and possibly other phenols - may alter the metabolic activity of bacteria in a way that allows them to mitigate any deleterious effects they might otherwise experience. In any case, these would be questions for future research and are not within the scope of the results acquired in this study.

ACKNOWLEDGEMENTS

We would like to express our sincere gratitude to our microbiology professor, Dr Catalina Cuellar-Gempeler, for her continued encouragement of our project and indispensable support with the process of revising our paper and communicating our findings. We would also like to thank our lab instructor, Gavin Sutter, and lab supervisor, Dejah Thrift, for working with our group to develop and carry out our experimental design. Lastly, we would like to thank Dr Claire Till for providing additional feedback and help with determining our statistical methods.

LITERATURE CITED

1. Boy FR, Benito MJ, Córdoba MdG, Rodríguez A, Casquete R. 2023. Antimicrobial Properties of Essential Oils Obtained from Autochthonous Aromatic Plants, *International Journal of Environmental Research and Public Health* 20, 1657-1667.
2. Ribinskas T, Vitkauskiene A, Kareiviene V, Zevzikoviene A. 2024. Antimicrobial Activity of Euphorbia

helioscopia L. Against Methicillin-Resistant Staphylococcus aureus (MRSA) In Vitro, *Cureus* 16

3. Silva O, Duarte A, Pimentel M, Viegas S, Barroso H, Machado J, Pires I, Cabrita J, Gomes E. 1997. Antimicrobial activity of Terminalia macroptera root, *Journal of Ethnopharmacology* 57, 203-207.
4. Puupponen-Pimiä R, Nohynek L, Meier C, Kähkönen M, Heinonen M, Hopia A, Oksman-Caldentey KM. 2001. Antimicrobial properties of phenolic compounds from berries, *Journal of Applied Microbiology* 90, 494–507.
5. Kujawska M, Jodynis-Liebert J. 2020. Potential of the ellagic acid-derived gut microbiota metabolite - Urolithin A in gastrointestinal protection, *World journal of gastroenterology* 26, 3170–3181.
6. Landete JM. 2011. Ellagitannins, ellagic acid and their derived metabolites: A review about source, metabolism, functions and health, *Food Research International* 44, 1150-1160.
7. Iglesias-Aguirre CE, García-Villalba R, Beltrán D, Frutos-Lisón MD, Espín JC, Tomás-Barberán FA, Selma MV. 2023. Gut Bacteria Involved in Ellagic Acid Metabolism To Yield Human Urolithin Metabotypes Revealed. *Journal of Agricultural and Food Chemistry* 71, 4029-4035.
8. Nobo S. 2017. Antimicrobial Efficacy of the Phenolic Compounds Ellagic Acid and Epigallocatechin Gallate, *Journal of McPherson College Science* 25, 42.
9. Skoog DA, West DM, Holler FJ, Crouch SR. 2014. Comparison of two experimental means. In *Fundamentals of Analytical Chemistry* (9th ed., p.134). Brooks/Cole Cengage learning. Chapter 7B-2.
10. Sax, J. K. 2015. Dietary Supplements are Not all Safe and Not all Food: How the Low Cost of Dietary Supplements Preys on the Consumer, *American Journal of Law & Medicine*, 41, 374–394.

Made in the USA
Coppell, TX
04 November 2026

11286706R00031